**The Natural History
and Behavior
of North American
Beewolves**

The Natural History and Behavior of North American Beewolves

Howard E. Evans
Department of Entomology
Colorado State University

Kevin M. O'Neill
Department of Entomology
Montana State University

Comstock Publishing Associates
a division of Cornell University Press
Ithaca and London

Library of Congress Cataloging-in-Publication Data
Evans, Howard Ensign.
　The natural history and behavior of North American beewolves /
　Howard E. Evans, Kevin M. O'Neill.
　　　p.　cm.
　Bibliography: p.
　Includes index.
　ISBN 0-8014-1839-9 (alk. paper)
　ISBN 0-8014-9513-X (pbk. : alk. paper)
　1. Philanthus—North America.　I. O'Neill, Kevin M.　II. Title.
QL568.P5E86 1988
595.79—dc19　　　　　　　　　　　　　　　　　　　　　　87-25073

Copyright © 1988 by Cornell University

All rights reserved. Except for brief quotations in a review, this book, or parts thereof, must not be reproduced in any form without permission in writing from the publisher. For information, address Cornell University Press, 124 Roberts Place, Ithaca, New York 14850.

First published 1988 by Cornell University Press.
Comstock|Cornell Paperbacks edition first published 1988.

Printed in the United States of America

The paper in this book is acid-free and meets the guidelines for permanence and durability of the Committee on Production Guidelines for Book Longevity of the Council on Library Resources.

Contents

	Acknowledgments	vii
1	Introduction	1
2	Major Features of the Biology of Beewolves	8
3	Species of the *zebratus* Group	32
4	Species of the *gibbosus* Group	78
5	Species of the *pacificus* Group	115
6	Species of the *politus* Group	140
7	Other North American Species of *Philanthus*	165
8	A Brief Review of Eurasian Species	192
9	An Overview of Male Mating Strategies	198
10	Major Features of Nesting Behavior, with a Final Look at Beewolves	230
	References	261
	Index	273

Acknowledgments

Our studies have from time to time been aided by grants from the National Science Foundation (BNS 76-09319 and BNS 79-26655), Sigma Xi, and the University of Wyoming–National Park Service Research Center. We have been fortunate in having as friends and colleagues two persons who themselves have made important contributions to the study of *Philanthus:* John Alcock and Darryl Gwynne. Others who have assisted us in the field include Byron Alexander, Keith Christian, Mary Alice Evans, Robert Longair, Ruth O'Neill, Henry Thoenes, and William Rubink.

Assistance in identifying species of *Philanthus* was provided by R. M. Bohart of the University of California, Davis, and by G. F. Ferguson of Oregon State University. Dr. Bohart also identified some of the prey; other assistance in prey identification was provided by G. C. Eickwort, Cornell University, U. N. Lanham, University of Colorado, R. McGinley, Smithsonian Institution, and T. B. Mitchell, North Carolina State University.

HOWARD E. EVANS
KEVIN M. O'NEILL

Fort Collins, Colorado
Bozeman, Montana

The Natural History and Behavior of North American Beewolves

Chapter 1
Introduction

Beewolves are ground-nesting, solitary wasps belonging to the genus *Philanthus* of the family Sphecidae. They prey on bees and other Hymenoptera, which they place in nest cells as food for their larvae. When naming the genus, in 1790, J. C. Fabricius chose the Greek equivalent of "flower-lover," reflecting the fact that adults take their nourishment at the nectar of flowers (not a particularly distinctive feature, as many insects fuel themselves at flowers). A few years later, in 1799, P. A. Latreille named the best-known European species *apivorus* ("bee-eater"), though unfortunately this appropriate name cannot be used, having been predated by Fabricius's name *triangulum*. Latreille also provided early behavioral observations of *Philanthus* nesting along a road near Paris: descriptions of digging behavior and of the capture of bees on flowers and at hive entrances.

There were other studies of *Philanthus triangulum* during the 19th century, but it was perhaps J. H. Fabre's treatment of the species in his *Souvenirs entomologiques* (1891) that attracted most attention. Fabre's essay appeared in English translation by B. Miall in 1920, under the title "The Bee-eating Philanthus." The anecdotal accounts of Fabre and other early workers were complemented, beginning in 1932, by a classic series of articles by Niko Tinbergen and his colleagues, which demonstrated the usefulness of simple field experiments in elucidating the cues used by the wasps in homing and in prey capture. This research was summarized by Tinbergen in his widely read books *The Study of Instinct* (1951) and *Curious Naturalists* (1958). It was Tinbergen who first popularized the word *beewolf*, although the term *bee pirate* may have a prior origin (Mally, 1909; Betts, 1936). In recent years the beewolf of Europe and Africa, *Philanthus triangulum*, has attracted the attention of several excellent researchers, foremost among whom are W. Rathmayer (1962a, 1962b) and R. T. Simonthomas and his colleagues (1966, 1972, 1978).

Philanthus is a genus of some 136 species, which occur on all continents except Australia, South America, and Antarctica. In the American tropics it is replaced by the closely related derived genus *Trachypus* (see Evans and Matthews, 1973a; Rubio, 1975). In Australia the beewolf role has been assumed by certain species of the genus *Bembix* (Evans and Matthews, 1973b). *Philanthus* is best represented in Africa, where (if one includes the Mediterranean area) nearly half the species occur. There are also about 30 species in Eurasia and the East Indies. Bohart and Menke (1976) have indicated that the more generalized genera of the subfamily Philanthinae, *Eremiasphecium* and *Philanthinus*, are confined to the Eastern Hemisphere, so it seems reasonable to assume that *Philanthus* had its origin there.

Philanthus is one of 11 genera included in the subfamily Philanthinae; 7 of these genera occur in North America (Table 1-1). In terms of size, the Philanthinae rank fourth among the 11 subfamilies recognized by Bohart and Menke (1976), containing about 1100 of the approximately 7700 known species of Sphecidae. Collectively, Sphecidae prey on members of nearly all orders of insects, as well as spiders, but only a few genera outside the Philanthinae use Hymenoptera as prey. Although called "digger wasps," some Sphecidae nest in stems or in cavities in wood, and some build mud nests well above the ground. Good reviews of the structure and behavior of the Sphecidae as a whole are provided by Bohart and Menke (1976), and Iwata (1976) has reviewed many aspects of female behavior. The Sphecidae are usually conceived as making up the single family of the superfamily Sphecoidea. The close relationship of the sphecids to the bees has long been recognized, however; and if the two groups are placed in a single superfamily, the name Apoidea should be applied to it (Michener, 1986). Although bees provision their nests with pollen and nectar rather than paralyzed prey, there are parallels between solitary bees and digger wasps in several aspects of both male and female behavior.

In his "Phylogeny and Classification of the Aculeate Hymenoptera," Brothers (1975) placed the digger wasps and bees together on a major branch of his dendrogram, another branch including all the other groups of wasps (except Chrysidoidea). Brothers preferred to consider

Table 1-1. An overview of the Philanthinae[1]

Classification	Distribution	Prey
Eremiaspheciini		
Eremiasphecium	So. Eurasia, No. Africa	Unknown
Aphilanthopsini		
Philanthinus	So. Europe, No. Africa	Unknown
Aphilanthops	No. America	Queen ants[2] (Formica)
Clypeadon	Western No. America	Worker ants (Pogonomyrmex)
Listropygia	Southwestern United States	Worker ants (Pogonomyrmex)
Odontosphecini		
Odontosphex	So. America, NW Africa, SW Asia	Unknown
Pseudoscoliini		
Pseudoscolia	So. Eurasia, No. Africa	Bees
Philanthini		
Philanthus	Worldwide except Australia, So. America	Bees, wasps, other Hymenoptera
Trachypus	So. Texas to Argentina	Bees, wasps
Cercerini		
Cerceris	Worldwide	Beetles[3]
Eucerceris	No. America	Weevils

1. After Bohart and Menke (1976).
2. However, Evans (1977) reported bees as prey of A. hispidus in Baja California.
3. But several Eurasian and African species use Hymenoptera as prey.

most of the subfamilies of Sphecidae as families, as did Krombein (1979) in the *Catalog of Hymenoptera in America North of Mexico*. We prefer to follow Bohart and Menke (1976) in considering the Sphecidae a single family, because there are genera that, in their words, "cut across traditional subfamily lines" (p. 30) and blur any breaks that might be used to justify family ranking.

Within the Sphecidae, the Philanthinae stand as one of the more highly evolved groups. Bohart and Menke (1976) considered 30 features of adult morphology, ranking each subfamily from 0 to 3 according to the prevalence of the more specialized alternative of each feature. The Philanthinae ranked with the Nyssoninae, Larrinae, and Crabroninae as most advanced (Fig. 1-1). Evans (1959, p. 183) commented that "the larvae of the Philanthinae are among the most specialized of all digger wasps." Of 10 larval characters evaluated, the Philanthinae were regarded as specialized in all but one. The dense spinules covering the entire body, the terminal pseudopod, and the reduced galeae of the mouthparts are especially notable. Of the other subfamilies, the Nyssoninae appear to have the most larval features in common with the Philanthinae (Evans, 1964a).

A case can be made that the Philanthinae are also among the most advanced digger wasps with respect to behavior. Similar claims can be made for the Nyssoninae (sand wasps), whose ethology has been reviewed by Evans (1966c). Although members of both of these major subfamilies nest in the ground, they have evolved along quite different lines, and there are major differences in both male and female behavior between nyssonine sand wasps and Philanthinae.

True to their name, most sand wasps prefer highly friable, sandy soil, and in such soil each female makes a series of nests containing one or a few cells. Many higher Nyssoninae have reduced ocelli, extensively pale coloration, and other adaptations for exploiting hot and arid conditions such as those found in deserts and dunes. Many species are progressive provisioners, bringing in prey over several days as the larva grows,

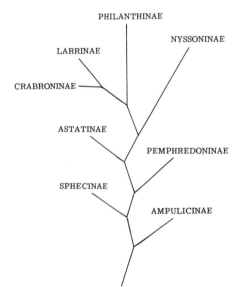

Figure 1-1. Relationships of the major subfamilies of Sphecidae (after Bohart and Menke, 1976). Three small subfamilies, each containing a single genus, have been omitted.

resulting in much mother-larva contact and a high survival rate among the limited number of offspring produced. Cocoon-spinning behavior of the mature larva is a remarkable process, involving incorporation of sand grains into the walls and construction of an elaborate series of pores. Presumably these cocoons permit the diapausing larva and pupa to survive if they are uncovered by wind or flooded by streams.

In many genera of sand wasps, males possess various cuticular processes on the legs and abdomen that evidently play a role in copulation. Males usually patrol the emergence area in numbers, resulting in what has often been described as a "sun dance"; but territorial defense has been described in a few species (e.g., in *Sphecius*; N. Lin, 1963).

Although Philanthinae also require soil that can be excavated with the limited tools available—mandibles, leg structures, and modifications of the terminal abdominal segments—they differ from sand wasps in that many species are able to exploit soil that is relatively coarse-grained and hard-packed, soil more often described as clay, silt, sandy gravel, or sandy loam. Related to their ability to use such difficult substrate is their tendency to make nests of many cells. The cells are constructed sequentially along short side burrows that branch from a more or less elongate main burrow, a method that results in less wear of irreplaceable body parts than would the building of many separate burrows. Philanthinae mass-provision each cell, then lay the egg and close the cell; thus there is no mother-larva contact. The cocoons contrast sharply with those of Nyssoninae, being delicate and thin-walled, tapered toward one end. Deep in relatively compact soil, they are presumably much less likely to be exposed to the elements than are those of Nyssoninae.

Male Philanthinae generally lack the cuticular processes of many male Nyssoninae, but nearly all male Philanthinae have brushes of hairs on the sides of the clypeus as well as an abundance of hairs on the underside of the abdomen. The clypeal brushes are associated with the aperture of mandibular glands that produce a sex pheromone; abdominal hairs apparently serve to spread the pheromone on vegetation on the periphery of defended territories (Chapter 2).

By no means are all Philanthinae beewolves, although most genera studied so far do show a predilection for Hymenoptera (Table 1-1). Our final chapters discuss some of these genera, as they shed light on behavioral trends within the genus *Philanthus*. Unfortunately, none of these genera have received the study they deserve.

Beewolves have much to recommend them as objects of study. They are among the most commonly encountered wasps and often occur in aggregations that persist from year to year. They have sometimes been reported from urban areas (Reinhard, 1924), and they occasionally raid the hives of domestic honey bees (more especially in Europe and Africa; see Chapter 2). There are 34 species in North America, more than half of which have been studied in the field. The behavior of both males and females is rich in details that vary from species to species, challenging the researcher to interpret these differences in terms of their adaptiveness to differing edaphic, climatic, and biotic conditions. But the ecological constraints on each species are not rigid; and furthermore, aggregations may be somewhat isolated from one another. Thus there is a further challenge to describe interpopulational differences as they relate to such matters as the abundance and patterning of available nesting substrate, the abundance of certain natural enemies, and prey availability.

Several years ago we set about to fill in as many details as possible toward the preparation of a broad-scale, comparative study of the behavior of the North American species of *Philanthus*. Several deficiencies in the resulting monograph are inherent in the nature of the subject. There are excellent published reports on several species, and we have found good populations of these species and several others. All species have certain geographical limitations, however, and all occur in localized populations of variable size; thus it was not always possible to obtain the kind and extent of information we wanted. Added to this difficulty is the fact that most species are active for only a few weeks of summer, so that data gathering comprises intensive episodes that cannot be renewed for many months. Laboratory studies of some aspects of female nesting behavior are possible (Simonthomas and Veenendaal, 1978); but by and large, meaningful comparative studies must be done in the field, where ecological factors as well as the spacing of nests and territories can be evaluated. Because of the prevalence of intraspecific variation in both male and female behavior, data should be obtained from several aggregations of each species whenever possible.

We have reviewed published information on North American *Philanthus* and have added a good deal that has not previously been in print. The available data are understandably uneven both in quantity and in quality. Nevertheless, we believe that enough progress has been made to justify a stocktaking and some preliminary conclusions concerning the evolution of behavior in the genus and the factors that appear to have shaped that evolution.

Our field techniques have been primarily observational, though occasionally supplemented with manipulation of the wasps' behavior. Observational methods included ad libitum sampling of behaviors such as aggressive interactions, matings, and predations, as well as focal samples of individuals or small groups of wasps at territories or nest sites. In many studies, we took regular censuses to record all active and visible males and females to obtain information on daily and seasonal patterns of activity. We collected samples of wasps, prey, and natural enemies for identification or dissection and excavated representative nests to obtain information on prey and nest structure. Maps of nesting and territorial areas were prepared for many species. To aid our observations, we marked nests and territories with stakes and marked wasps on the dorsum of the thorax with one or more dots of enamel paint to permit individual or size-class recognition. In the field, to indicate body size, we measured head width to the nearest 0.1 mm with a VWR Scientific Products micrometer accurate to 0.05 mm. Head width was used as an indicator of body size because it is stable (unlike body length, which varies because of the telescoping abdomen), it is easily measured, and it correlates strongly with body mass and with other linear measures (O'Neill, 1983a, 1983b, 1985).

One of our major techniques for studying territorial behavior of males involved *removal experiments* (O'Neill, 1983a, 1983b). These experiments allowed us to obtain information on the relative body sizes of territorial and nonterritorial males and to estimate the relative numbers of males in both groups. The experiments were conducted as follows: (1) a resident male was removed from his territory, measured, and held captive in an insect net; (2) if he was replaced on the territory within 30 minutes by another male who also began defending the area, this second male (the first replacer) was also removed and measured; and (3) in some cases, the experiment was continued until no further males entered and defended the territory (all males beyond the first replacer are termed *subsequent replacers*).

We first consider, in Chapter 2, the major features of behavior common to all or most species of *Philanthus*. In this section we rely heavily on the extensive research that has been conducted on *triangulum*. We also use Chapter 2 to clarify the terminology to be employed in the descriptions of individual species in Chapter 3–7. These species are arranged for the most part according to the accepted classification into species-groups (Bohart and Grissell, 1975). A brief review of the major behavioral features of several species of the Eastern Hemisphere follows (Chapter 8). Two final chapters represent an effort to synthesize information derived from comparative studies and to draw conclusions concerning the comparative ethology and evolution of these wasps.

Chapter 2

Major Features of the Biology of Beewolves

Studies of the life histories and behavior of species of *Philanthus* suggest that certain features are common to most, if not all, species. This chapter reviews these features so that they need not be repeated under each species. It also reviews certain aspects of behavior that have been well studied in only one species (often *triangulum*) but are very probably characteristic of all species, at least in a general way.

A few remarks on structure and color also seem in order. All species show sexual dimorphism to varying degrees, females invariably larger on average than males. O'Neill (1985) summarized size data on four species of *Philanthus*; in every case females were significantly larger than males. Size in females correlates well with size of their ovarial eggs and size of the prey they provision. Females also need space for the musculature of the mandibles, front legs, and wings, all of which play important roles in nest digging and in carrying prey in flight. These features evidently provide the selective pressure that results in the evolution of sexual size dimorphism in beewolves and many other wasps.

Males of the majority of species have a series of transverse yellow bands on the gaster and often various bands and spots on the head and thorax, approaching a pattern that is mostly yellow in a few species (e.g., *multimaculatus, parkeri*). Females tend to have somewhat less extensive maculations, often paler in color and more complex in pattern. Sexual dimorphism reaches an extreme in *bicinctus*, in which the first tergite of the female gaster has an orange band, the second has a yellow band, and the remainder of the gaster is black (in contrast to the six yellow bands on the gaster of the male).

Sexual color dimorphism occurs among digger wasps of many groups. Color patterns may play a role in sex and species recognition or in thermoregulation. The reasons color dimorphism is so marked in species such as *bicinctus* and much less marked in other species (e.g., *barbatus*) remain obscure. In one respect there is less dimorphism in *Philanthus* than in many other genera: both sexes have a well-formed pecten (also called tarsal comb or rake) on the front legs. Males are reasonably effective diggers, although the spines of the male pecten, being shorter and thinner than those of females, scarcely provide the equipment needed to dig deep nests like those the females dig.

Males also differ from females in having a pair of hair tufts, resembling small brushes, on each side of the clypeus (an exception is *albopilosus*). These hair tufts are associated with large mandibular glands in the head, which open at the base of the brushes and are believed to produce a sex pheromone, the clypeal brushes serving as applicators (Gwynne, 1978; Schmidt et al., 1985). There are also, in males of most species, numerous hairs on the ventral side of the gaster; they are not associated with any known glands and appear to act as a

Figure 2-1. A male *Philanthus bicinctus* on a territorial perch. The abdominal hair brushes are conspicuous. A numbered disc attached to the mesoscutum identifies this male. (Photograph by Darryl T. Gwynne.)

means of spreading the secretions of the mandibular glands after the secretions have been applied to stems or leaves by the clypeal brushes. In several species these hairs are dense and quite long, varying from amber to black (e.g., in species of the *zebratus* group, Chapter 3) (Fig. 2-1). Other species have hairs that are abundant but somewhat shorter and paler (e.g., in members of the *gibbosus*, *pacificus*, and *politus* groups). Some species have sparse hairs (*triangulum*, *bilunatus*), others only a very few, short hairs (*solivagus*, *lepidus*, *albopilosus*). In *crabroniformis*, a species with dense gastral hair brushes, a fringe of close-set hairs appears along the apical margin of the second sternite. The abundance, length, and patterning of these hairs may bear some subtle relationship to male reproductive behavior, but this relationship remains to be explored.

Habitat and Life History

Because all species of beewolves are ground nesters, they tend to occur locally where the soil is moderately to strongly friable, either fine-grained sand or silt or more consolidated clay, clay-loam, or sandy loam. Even a soil whose surface is stony, encrusted, or compacted (as in trails and dirt roads) may serve as suitable substrate for some species if sufficiently friable beneath the surface. Most species prefer flat or slightly sloping soil, but several nest in flat soil, steep slopes, or vertical banks (*gibbosus*, *solivagus*); one species (*inversus*) has been found nesting only in steep to vertical banks. Males typically spend the night and periods of inclement weather in the nesting area of the females; during the day they tend to remain within at least a few hundred meters of this area, usually much closer than that.

In temperate climates, all species overwinter as diapausing larvae (prepupae) in cocoons in the soil. The larvae of several species have been described and figured by Grandi (1961) and by Evans (1957a, 1964c). Pupation occurs in spring or summer, presumably in response to warming of the soil, and adults emerge a few weeks later. In a given area, emergence of species tends to be staggered, one or more species appearing early in the summer, others in midsummer, others in late summer (Evans, 1970). Most populations are univoltine, but bivoltine populations have been described for some species; and where the summer season is long, some species may breed continually through the summer.

Protandry (emergence of males before females) is common, perhaps universal. Gwynne (1980) reported a sex ratio close to unity, both on the basis of observing emergence into a cage in the field and on the basis of marking large numbers of individuals of both sexes. Other reports support the belief that sex ratios close to 1:1 are usual. Male beewolves are relatively long-lived. Evans (1973) recorded males of *gibbosus* as living up to 49 days from the time they were first marked. Simonthomas and Poorter (1972) reported a lifespan of 33 days for males compared with 38 days for females. The long life of males seems curious because evidence suggests that in beewolves and many other wasps the females mate only once; this practice is suggested by the infrequency of observed copulations and the frequency of female refusals (Alcock et al., 1978). Females do not necessarily mate immediately after emerging but may instead do so after nests have been initiated (Gwynne, 1980). The continued presence of males does, of course, enable them to mate with late-emerging females.

The daily regimen at an aggregation of beewolves is somewhat as follows. Females tend to emerge from the burrows earlier than males, often between 0830 and 1000 hours, depending on temperature and insolation. Members of either sex may remain in the entrance facing out for a time before emerging. Females clear the burrow and entrance of soil for a period in the morning (or initiate a new nest at this time). First flights from the nest for prey may occur in late morning and continue at intervals until late afternoon. Males undertake patrolling flights or establish territories in mid to late morning, but most male activity has declined by midafternoon. Members of both sexes visit flowers for nectar periodically, but especially in the morning and late afternoon.

Male Behavior

Males commonly make shallow burrows in or near the nesting area of the females, where they spend the night and periods of inclement weather (Figs. 3-11, 6-6F). The burrows are closed from the inside when occupied but left open when the males are outside; each usually has a small mound of soil outside the entrance. Males may reoccupy the same burrow on successive nights, and at times more than one occupies the same burrow. Males may also utilize other available holes, such as abandoned nests of wasps or bees. Sharing of active nests with conspecific females has been reported for some species.

Following their emergence in the morning, males fly about near the site of emergence, often perching on the ground or low plants and frequently visiting flowers for nectar. During this time they show no apparent site attachment and usually do not interact with conspecific males or females. After a time, which may vary from 30 minutes to 1 or 2 hours, depending on temperature and sunlight, males begin to show evidence of mate-seeking activities. At first such activities are likely to include occasional attempts to mate with females digging at nest entrances and aggressive interactions with other males. By mid to late morning most males are performing the reproductive strategy characteristic of their population and size class, as discussed further along.

Male territoriality has been described in 18 North American species of *Philanthus* and in the European species *triangulum*. A *territory* is defined as "a fixed area from which intruders are excluded by some combination of advertisement, threat, and attack" (Brown, 1975, p. 61). Territories of male *Philanthus* vary in diameter from about 0.5 to 1.5 m, although males sometimes respond to moving objects outside territorial boundaries and may pursue insects for several meters. Each territory includes a more or less central perch, usually in a relatively bare spot, with surrounding vegetation. Vegetation is from time to time scent-marked by means of behavior often described as "abdomen dragging" (Alcock, 1975c, 1975d). The male lands on a stem and walks up a short distance, then down again over a similar distance, pressing his head and abdomen against the stem, assuming a broadly V-shaped posture (Fig. 2-2). Most persons who have studied this behavior have assumed that a pheromone is being applied from mandibular glands through the clypeal brushes (Figs. 2-3, 2-4) and spread by abdominal hair brushes, thus facilitating evaporation of the chemical (Figs. 2-1, 9-1). No glands have been found associated with the abdominal hairs. Males tend to do much scent marking soon after they have established a territory but often mark less frequently as time progresses. An individual flight terminating in scent marking we term a *bout*; bouts may consist of the marking of only one stem or of several stems in rapid succession.

Females of all *Philanthus* species, as well as males of the one species, *albopilosus*, in which males are never territorial, possess relatively small mandibular glands (Fig. 2-4). In males of other species, a duct emanating from each gland apparently exits near the base of the mandibles at the point where the clypeal hair brushes emerge (Gwynne, 1978). Close examination of the base of these hairs reveals a small hole where each hair articulates with the clypeus (Fig. 2-3, right). The clypeal hairs seem ideally structured to act as brushes for applying chemicals to a substrate: their longitudinal grooves may act to move the secretion along the hairs by capillary action. On newly emerged males

Figure 2-2. A male *Philanthus crabroniformis* scent-marking a grass stem. The ventral surfaces of the head and abdomen are applied closely to the stem, the body assuming a broadly V-shaped posture.

and those preserved in alcohol, the hairs are well separated. In contrast, on males collected on territories, the hairs in the brush adhere to one another, as if cemented with a secretion. As detailed earlier, when a male walks along a stem in his territory, the mandibular area is closely appressed to the surface of the plant.

In *triangulum* (Borg-Karlson and Tengö, 1980) and *crabroniformis*, as well as *Eucerceris arenaria*, an odor can be detected emanating from plant stems on which males have recently abdomen-dragged. In many species, no such odor is apparent; however, more direct evidence for a scent-marking function is now available. Extracts of stems freshly marked by abdomen-dragging males of *basilaris* contained all the volatiles found in the mandibular glands of males (Schmidt et al., 1985). Control stems contained none of these chemicals. Thus there is little doubt that territorial males of *Philanthus* deposit substances from the mandibular glands onto stems within their territories. Vinson et al. (1982) also reported a one-to-one correspondence between grass and head extracts for territorial males of the bee *Centris adani*.

The chemistry of mandibular gland secretions has been studied in three species of *Philanthus* (*basilaris*, *bicinctus*, and *triangulum*). Schmidt et al. (1985) and McDaniel et al. (1987) found large quantities of

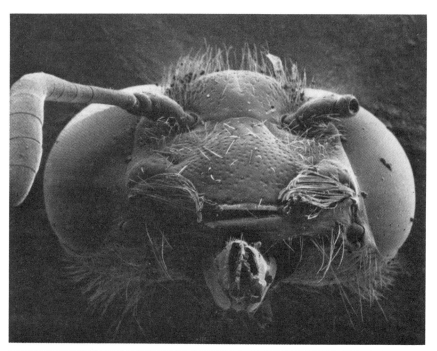

Figure 2-3. Scanning electron micrograph of the head of a male *Philanthus gibbosus*, showing the clypeal brushes (top) and an enlargement (× 175) of the base of the clypeal hairs (bottom).

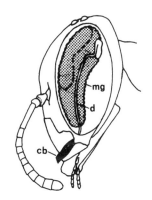

Figure 2-4. Lateral view of the head of a male *Philanthus psyche* with the compound eye cut away to show the size and position of the mandibular glands (*mg*; entire stippled area). The position of a duct (*d*) apparently leading from the mandibular gland to the area near the base of the clypeal brush (*cb*) is included, following Gwynne's (1978) dissection of *P. bicinctus*. The smaller area within the mandibular gland marked off with a dashed line shows the relative size of the mandibular gland of *P. albopilosus*, a species that does not scent-mark (clypeal brushes are also absent in this species).

volatiles in the mandibular glands of *basilaris* and *bicinctus*: 220 μg and 240 μg per head, respectively. Heads of both species contain complex, species-specific mixtures of ketones, fatty acids, ethyl esters, and aldehydes, the major component for both species being 2-pentadecanone. All of the volatile components identified had molecular weights typical for insect sex pheromones (Wilson, 1970). The chemicals were absent in thoracic and abdominal extracts of males and in all extracts from females.

Several studies have been conducted on the cephalic volatiles produced by *triangulum* and have given different results. Borg-Karlson and Tengö (1980) reported that males in a Swedish population produced at least three types of pyrazines. However, there are reasons to believe that these substances may not be the marking volatiles. First, identical or related pyrazines were also found in the heads of female *triangulum* and in individuals of three species of nyssonine wasps (Borg-Karlson and Tengö, 1980), four species of *Ammophila* (Duffield et al., 1981), and five species of eumenine wasps (Hefetz and Batra, 1980), none of which are known to scent-mark. Second, the chemicals were present in relatively small amounts (0.5 μg per head) compared with the volatiles found in *basilaris* and *bicinctus*. Third, it was not confirmed that the chemicals were actually applied to the stems. Finally, a recent analysis of individuals of a population of *triangulum* from southern France gave quite different results (C. A. McDaniel, J. O. Schmidt, and R. T. Simonthomas, in preparation). Head extracts from males, but not females, in this population consisted primarily of Z-11-eicosen-1-ol, with smaller amounts of ketone, fatty acids, and aldehyde. This mixture was thus more closely related to the chemicals found in *basilaris* and *bicinctus* than to pyrazines. Whether the observed differences between the two populations of *triangulum* represent functional (or, for that matter, nonfunctional) geographic differences or simply differences in biochemical methodology (J. O. Schmidt, personal communication, 1987) remains to be seen.

Territories bear various spatial relationships to the female nests, depending at least in part on nest density. When nests are widely dispersed, as they often are in *basilaris*, for example, male territories may be grouped in *leks*, a lek being defined as an aggregation of competing males in an area containing no resources for females other than the males themselves.

Evidence suggests that certain territorial sites are superior to others, as indicated by the fact that they are more frequently occupied. Territorial sites and even individual territories have been seen to be occupied in successive years, of course by different generations of wasps. The same male may occupy one territory for several hours on each of several successive days, but that is more the exception than the rule. Not uncommonly a territory is occupied by several different males in the course of one day (*serial territoriality*, Alcock 1975d).

Territory residents engage in frequent interactions with other males, both those in adjacent territories and those moving about and attempting to occupy a superior territory. Such contests are frequent soon after territorial sites have been selected in the morning (Fig. 3-5). Contests are almost always settled in favor of the larger male, without respect to whether he happens to be the resident (O'Neill, 1983a). Interactions between males are of several kinds, which tend to grade into one another. In *triangulum*, Simonthomas and Poorter (1972) described a sequence in which the resident pursues the intruder and then the pair circles one another, the resident taking a higher position. This sequence may be followed by "a downward zigzagging swinging flight by which the intruder is re-encountered" (p. 144). Sometimes these flights end with the resident perching and the intruder perching nearby. These authors never observed physical contact between skirmishing males.

Alcock (1975b) discerned eight different behavior patterns performed by interacting males of *multimaculatus*.

1. Strike and flee: intruder approached perched male, struck him by dropping onto his back, then flew off.
2. Hover-orient: resident left his perch after having been struck or on the approach of another male, then hovered while oriented toward the intruder.

3. Darting flight: following hover-orient, resident often flew rapidly at the side of his rival, but contacts were rare.
4. Swirling flight: pair flew side-by-side in circles and loops with such speed that it was difficult to follow them visually.
5. Vertical flight: a male flew slowly upward with the other wasp trailing under and slightly behind it at a distance of 2–4 cm.
6. Drop-strike: the higher of the two males in vertical flight dropped onto the back of the other male before reaching a height of 50 cm. Sometimes one or both males were driven to the ground.
7. Grappling: males fell to the ground and grappled briefly.
8. Pursuit flight: one male flew straight off with the other directly behind it.

In his studies of *bicinctus*, Gwynne (1978) recognized four categories of interactions of increasing intensity.

1. Approach: resident flies from his perch in direction of intruder.
2. Butt: resident makes physical contact with intruder but does not grasp intruder.
3. Grapple: male grasps the intruder but the pair does not fall to ground.
4. Grapple to ground: grappling pair falls to ground.

The term *butt* implies a strike with the head, sometimes audible, whereas the term *grapple* implies use of the mandibles and legs in a physical struggle. Some of the described interactions probably result from a male's response to what he believes to be a female; these include some instances of grappling as well as "strike and flee" (type 1 of Alcock). Indeed, except in the case of prolonged aggressive interactions between males, it is sometimes difficult to distinguish between aggressive and investigative behaviors. Overall, one may say that species differ in the type, the frequency, and the intensity of male interactions.

Exclusion of males from territories does not necessarily mean that they are excluded from the mating system. Such males may fly from territory to territory, ready to occupy territories that are less favorable or that have been vacated by other males. Such males have been termed *visitors* (Simonthomas and Poorter, 1972), or *floaters* (O'Neill, 1983a). Males perched between or adjacent to territories may be spoken of as *satellite males*. There is evidence that such males do sometimes mate successfully (see Chapter 5, under *pulcher*).

It appears that many species of *Philanthus* exhibit dual tactics, territoriality being the major tactic, performed primarily by larger males. Smaller males act as floaters or satellites. Certain species provide exceptions to this statement, however. In *zebratus*, larger males participate in a high patrolling flight, whereas smaller males are territorial (Chapter 3); and in *albopilosus* no evidence of male territoriality has been discovered (Chapter 6). Although in all species both sexes visit flowers for nectar, and females visit them for prey, there is no evidence that mating is initiated on flowers. Nor is there evidence of mating inside the burrows, as was suggested long ago by Reinhard (1924).

Copulations are seen infrequently. Simonthomas and Poorter (1972) observed only 3 in *triangulum*, Gwynne (1980) only 13 over a 3-year period in *bicinctus*, O'Neill and Evans (1983a) only 11 over 4 years in *zebratus*. The paucity of observed matings may occur because females mate only once, but it is frustrating to researchers wishing to confirm the success of differing male reproductive tactics. In general, males descend on females from above and grasp them while facing the same direction; very shortly both turn to face in opposite directions (Fig. 2-5).

Figure 2-5. A copulating pair of *Philanthus bicinctus*, the female above the male. The sexual color dimorphism is clearly evident in this photograph. (Photograph by Darryl T. Gwynne.)

Mating lasts several minutes, during which the pair may fly a short distance one or more times.

Female Behavior

Within one or a few days after they emerge, females dig shallow burrows here and there before remaining in one place to dig persistently. Digging may be discontinued from time to time while the wasp fuels herself at nectar sources. The mandibles are used for loosening the soil and for dragging out pebbles or other objects in the burrow (Olberg, 1953, pp. 44, 45). Soil is scraped from the burrow with synchronous strokes of the two front legs, which are curved toward the midline of the body so that the rake spines are brought into play. Scraping of soil is accompanied by rapid up-and-down movements of the abdomen. Periodically the wasp backs quickly from the burrow several centimeters and works forward, scraping sand behind her. As the burrow deepens, the female is seen less frequently at the entrance. The result of digging behavior is a pile of soil outside the entrance, called the *mound* (or *tumulus*). In some species, the mound is leveled more or less completely; the female backs to the far end and works forward with some zigzagging, scraping soil and dispersing it widely. Leveling movements may be interspersed with the later stages of digging or may follow completion of the initial, oblique burrow.

To make the initial and later closures of the nest, the female scrapes soil over the entrance from several directions. Very commonly, after the initial closure, she opens and reenters the nest, then repeats the closure; this reopening and reclosing may occur as many as four times before she leaves the nest. Females of some species dig one or more short burrows within 1 to 3 cm of the true nest entrance and at an angle to it (Figs. 2-6, 2-11). These *accessory burrows* (or *false burrows*) may serve as a source of fill for the closure, but they evidently also serve to divert the attention of hole-searching parasites (Evans, 1966a; Edmunds, 1974). Accessory burrows are not normally closed and may occasionally be reentered or expanded by the wasp.

On the first flight from the nest, the female makes an *orientation flight* (or *locality study*). This flight consists of various loops and figure eights over the entrance, often facing the entrance. In some species the flight is performed rather close to the ground, whereas in others the wasp ascends several meters before flying off; the altitude of the flight appears to correspond to the height at which the female returns to the

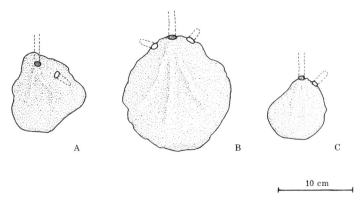

Figure 2-6. Typical tumuli and accessory burrows of three species of *Philanthus*: (A) *gibbosus*, (B) *zebratus*, (C) *crabroniformis*. See also Figures 2-7, 2-8, and 2-11.

nest with prey. Orientation flights are repeated each morning at the first departure from the nest and may be repeated on subsequent departures, usually in abbreviated form unless there has been some disturbance to the nest area. During the flight the female learns the nest position in relation to various landmarks. Generally, a flight of only a few seconds is sufficient to ensure prompt location of the nest upon return. Even in dense aggregations, females almost invariably find their nest entrance quickly, even though the entrance has been closed and (in some cases) the mound of soil at the entrance fully leveled.

Distance orientation is not well understood but evidently involves learning and recognition of large objects distant from the nest, such as trees. By a series of elegant experiments, Tinbergen and his coworkers showed that objects close to the nest differ greatly in their quality as landmarks (Tinbergen, 1932; Tinbergen and Kruyt, 1938; Beusekom, 1948; available in part in English in Tinbergen, 1972). These experiments were performed with *triangulum* but surely apply in a general way to all species of the genus. The wasps prefer to use (1) three-dimensional over flat objects, (2) patterned over uniformly colored flat objects, (3) large over small objects, (4) objects contrasting with the background over those matching its color, and (5) objects present at the first departure over those added later.

Nest Structure

Like most digger wasp females, each female beewolf, acting alone, digs a nest that provides her developing offspring with both a suitable

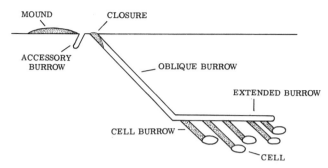

Figure 2-7. Sketch of a generalized *Philanthus* nest, showing the terminology used.

physical environment and protection from predators and parasites. Some of the terms used to describe nests are shown in Figure 2-7. The initial nest burrow is oblique, forming a 20–50° angle with a flat surface; in a slope the angle may be greater than this, and in a vertical bank it may form a 90° angle with the slope. We nevertheless call this initial burrow the *oblique burrow*. No cells are constructed from this burrow. Rather, it is extended in one of several ways. It may continue more or less directly downward or obliquely at a steep angle, with cells constructed sequentially in a diffuse, somewhat radial pattern from the bottom of this extension. Such nests we term *declinate* (bent downward). These are typically short-term nests; after one to five cells have been constructed, provisioned, and closed, the burrow is filled and another, similar nest is made (Fig. 5-4).

In other types of nests, the burrow is extended beyond the oblique section at a gentle forward slope or horizontally. We speak of these nests as *proclinate* (bent forward) (Figs. 3-16, 4-5, 7-1A). Such nests are typically more permanent than declinate nests and eventually may contain numerous cells; sometimes a female makes only one such nest, which serves her for the duration of her life. Cells are constructed at the ends of short *cell burrows* (Evans, 1957b) that branch from this *extended burrow*; each cell burrow is packed with soil after it has been fully provisioned and the egg laid. The extended burrow often passes up and down several times and may have sharp bends and two or more branches. The pattern of cells is often diffuse and irregular. The first cells may be constructed deep in the soil and additional cells added closer to the entrance; such nests are said to be *regressive* (Iwata, 1976) (Nielsen, 1933, spoke of such nests as *branché d'en bas*). Alternatively, cells may be constructed progressively farther from the entrance as the burrow is lengthened (*progressive*; Iwata, 1976) (in Nielsen's terminology, *branché d'en haut*). Some progressive nests have the cells constructed in a fairly regular series, one on each side of the burrow; such nests are often described as *serial* (Figs. 2-7, 3-2, 4-2). Serial nests have been known to be reoccupied and expanded by a second generation of wasps (see especially *gibbosus*, Chapter 4).

Although in general these terms are useful in describing nest structure, allowance must sometimes be made for intraspecific variation. Nests of *gibbosus*, for example, are usually serial, but when females nest in hard-packed substrate the cells are constructed in a more irregular pattern. Such variation may account for the lack of agreement in published observations on certain species.

To summarize, we recognize the following nest types:

>Declinate (short-term; diffuse cell pattern)
>Proclinate (long-term)
>>Regressive (diffuse cell pattern)
>>Progressive
>>>Diffuse cell pattern
>>>Serial cell pattern

Provisioning

The prey of beewolves consists of hymenopterans (usually bees and wasps) of sizes appropriate to each species of *Philanthus*. As a rule, both male and female bees and wasps serve as prey, except when the prey consists of social species, when only workers are used (workers and males, in the case of bumble bees). Many species are quite unselective of prey; others confine themselves mainly to bees of one family (usually Halictidae), and a few appear to specialize on members of one genus (*bicinctus* on *Bombus*, *inversus* on male *Agapostemon*) or even one species (*triangulum* on *Apis mellifera*).

Prey is grasped primarily by the wasp's middle legs during flight to the nest (Fig. 2-8). Entry into the nest may be direct, the wasp using her front legs for scraping open the entrance, or the prey (especially if rather large) may be deposited outside the entrance and drawn in from inside the burrow. In the latter instance, the antenna or one of the legs of the prey is grasped in the mandibles. When adjusting the prey at the nest entrance, and also sometimes in flight, the wasp may also use her mandibles (as well as her legs) to grasp the prey.

In some instances, ground-nesting bees may be captured outside or

Major Features of the Biology of Beewolves 23

Figure 2-8. A female *Philanthus crabroniformis* entering her nest, holding a small bee tightly beneath her abdomen with her middle legs. The front legs are being used to open the burrow, while the hind legs support the body laterally. An open accessory burrow is visible to the right.

inside their burrows (described under *gibbosus* and *crabroniformis*, Chapter 4), and honey bees may be taken at hive entrances (described under *sanbornii*, Chapter 3; *crabroniformis*, Chapter 4; and *triangulum*, Chapter 8). Much more commonly, prey capture occurs on flowers. The mode of hunting on flowers has been studied by Tinbergen (1935). Flying into a patch of flowers being visited by bees, a female first orients visually to a flying object of appropriate size. She then hovers downwind of the object and, if a suitable odor stimulus is received, flies and seizes it. Stinging occurs, however, only upon receipt of certain visual and tactile stimuli. The act of stinging has been studied by Rathmayer (1962a, 1962b), who found that the wasp attacks the bee from the front, grasping the bee's thorax from above and curving its abdomen beneath the bee. A single sting is made on the underside of the bee behind the front legs. Sensilla on the sting sheath appear to guide the sting to the membranous area behind the front coxae. The bee frequently attempts to sting the wasp, but in this position its sting contacts the sclerotized abdomen of the wasp, which it cannot pierce. The wasp's poison spreads first around the puncture in the bee and then, by means of the hemolymph, to the thoracic musculature. Paralysis soon spreads to the entire body. Contractions of the heart can still be observed after 37 hours, and some reports suggest that some bees may live several days, though in deep paralysis (Hamm and Richards, 1930). Rathmayer (1966) induced beewolves to sting honey bees in abnormal parts of their bodies by cutting holes in the bees' integument and placing the sting there. The farther the sting was inserted from the muscles of the legs and wings, the longer it took for movements to cease.

Tinbergen (1935) found that *triangulum* females will orient visually to flies, but unless the scent of honey bees is present, no prey is seized. Fahringer (1922) substituted drone flies (*Eristalis*), which are visual mimics of honey bees, for honey bees at nest entrances, but they were never accepted. Armitage (1965) observed *bicinctus* females grasping flies that mimicked the normal bumble bee prey but quickly releasing them. We have never found bee or wasp-mimicking flies in any of the many nests we have excavated. However, Stubblefield (1983) reported that among 2394 items of prey taken from *sanbornii* females, there was one fly, a stratiomyid of the genus *Odontomyia*. Gambino (1985) also reported a syrphid fly, *Sphaerophoria cylindrica*, from a nest of *neomexicanus*. Both of these are yellow-banded flies that may be regarded as wasp mimics.

Rathmayer (1962a) found that the venom of *Philanthus triangulum* is effective against insects of many kinds, including closely related Sphecidae. His research indicates that *Philanthus* is immune to its own poisons, toxicity evidently being destroyed by a substance in the hemolymph. O'Neill and Evans (1981), however, found that female *basilaris* sometimes prey on territorial males of their own species as well as on males of *bicinctus*.

Prey is not usually taken directly to a cell but is allowed to accumulate in the burrow at various points, usually at the bottom of the oblique burrow. Only later is the necessary cell burrow made and a cell constructed. Simonthomas and Veenendaal (1978) induced *triangulum* females to nest between sheets of glass so that behavior inside the nest could be studied. They found that females deposit prey in the burrow, then move to the end of the burrow and extend it, moving soil behind them. The bees are then dragged farther down, through the accumulated soil. This procedure is carried out in several stages and may require several hours. At times the bees are squeezed and the nectar licked up. Eventually, a cell burrow is dug at a right angle to the main burrow, and a cell is constructed at the end of it. The bees are dragged to the entrance of the cell, where they are squeezed for nectar once more. They are then fitted into the cell and removed again. The wasp reenters the cell and appears to measure it. She then replaces the bees, after several pauses and further squeezing of the prey. Oviposition follows on the bee nearest the entrance of the cell.

Malaxation (squeezing) of the prey has often been reported in *triangulum* and was the major theme of Fabre's (1891) essay on this species. Fabre maintained that bees are sometimes used solely for the nectar that can be squeezed from their crop, after which they are discarded. He also believed that nectar might be repugnant to the larvae and was therefore removed from the crop before the bees were placed in a cell. It is odd that this behavior has not been observed for any of the North American species of *Philanthus*. However, Kurczewski and Miller (1983) reported that even though many bees captured by *sanbornii* females bore pollen, none of the bees taken from cells had pollen on their bodies. They surmised that the wasps clean the bees and seal off the pollen at the end of a side burrow. They found one such cache of pollen that might have represented the product of cleaning behavior.

Weight of the cell contents considerably exceeds that of the adult wasps, as would be expected, because the larvae do not consume the exoskeletons of the prey. Kurczewski and Miller (1983) found that the total weight of bees in individual cells of *sanbornii* varied from 241 to 465 mg, whereas a female wasp weighed 134 mg. There is evidence that some (usually larger) cells contain more prey than others and that these more abundantly provisioned cells produce females. Simonthomas (1966) reported that of over 500 *triangulum* cells excavated, 19% contained one honey bee, 41% two bees, 22% three bees, 10% four, 7% five, and 1% six. Simonthomas and Veenendaal (1978, p. 7) stated that: "Generally speaking one or two bees are put into a breeding cell with an unfertilized egg which will develop into a male wasp. A fertilized egg which will produce a female wasp is provided with three to five bees."

It is not uncommon to find dead bees lying outside the nest entrances of *Philanthus* wasps, evidently having been dragged from the nest by the female (Fig. 3-3). Simonthomas and Simonthomas (1972) noted this many times in *triangulum* and often saw females remove bees during digging periods. They found that about one-third proved to have been parasitized by flies.

The egg is laid on the topmost prey in the cell, which lies venter-up. The egg is somewhat loosely attached, with the anterior end near the fore coxae, the egg extending backward longitudinally or somewhat obliquely along the body (Fig. 2-9). This is oviposition type 15 of Iwata

Figure 2-9. The egg of *Philanthus gibbosus* attached to the ventral surface of a halictid bee.

(1976), who pointed out that it is characteristic of at least three other genera of Philanthinae (*Aphilanthops*, *Cerceris*, and *Eucerceris*). Following oviposition, the female closes the cell burrow with a firm plug of soil and does not revisit the cell. Development is rapid; the egg hatches in only about 2 days, and larval development requires no more than a week before the cocoon is spun.

Natural Enemies

Certain aspects of behavior appear to have evolved in response to the pressure of predators and parasitoids. Chapter 10 explores these behavioral adaptations, but we consider natural enemies briefly here with respect to their mode of attack.

Predators on Adults

Olberg (1953) reported crab spiders (Thomisidae) waiting in flowers and preying on *triangulum* adults as they sought nectar, and there are records of *gibbosus* falling prey to crab spiders (Chapter 4). Gwynne has found lycosid spiders preying on female and male *bicinctus* in the nesting area (see Chapter 3). As mentioned above, *basilaris* females have been observed using male *basilaris* and *bicinctus* as prey.

Gwynne and O'Neill (1980) reported predation by robber flies (Asilidae) on four species of *Philanthus* (Fig. 2-10). Altogether, there were 17 records of predation on males and 2 on females. There is evidence that

Major Features of the Biology of Beewolves   27

Figure 2-10. An asilid fly (*Proctacanthus micans*) that has suspended itself from a twig and is feeding on a male *Philanthus bicinctus*. (Photograph by Darryl T. Gwynne.)

robber flies are attracted to places where males are territorial and may even be approached by males in the defense of their territories.

Parasitoids of Adults

Flies of the family Conopidae are occasionally seen perching on vegetation overlooking nesting sites. Conopids are reputed to pounce on adults and oviposit directly into the abdomen, the larva developing and eventually pupating in the abdomen, causing the death of the wasp. The genera associated with *Philanthus* (though with relatively little substantiating data) are *Physocephala* (Olberg, 1953) and *Zodion* (see under *zebratus* and *pulcher* in Chapters 3 and 5).

Parasitoids of Larvae

Bee flies (Bombyliidae) have occasionally been reported to attack *Philanthus* (see under *gibbosus*, Chapter 4). Flies of this group deposit eggs in open holes in the ground, the larvae working their way down to the cell and developing as external parasitoids of the larvae.

Velvet ants (Mutillidae) are commonly seen in *Philanthus* aggregations, but their presence there does not necessarily mean that they are associated with any particular host (various bees and wasps often nest in or near beewolf sites). There seem to be few authentic records of Mutillidae attacking *Philanthus* (see under *sanbornii*, Chapter 3, and *gibbosus*, Chapter 4). Mutillids enter open holes or dig through closures, penetrate to the cells, and deposit eggs on larvae (often after they have spun their cocoons).

Cleptoparasites

Cuckoo wasps (Chrysididae) have frequently been reported as being attracted to *Philanthus* nests, probably visually. They may enter open nests, dig through closures, or wait outside the nest and follow prey-laden females when they enter. Simonthomas and Simonthomas (1972) reported *Hedychrum intermedium* females attempting to attach an egg to the prey of *triangulum* females as they enter the nest. This observation suggests that this species may be a cleptoparasite, destroying the host egg when it is laid but developing primarily on the provisions in the cell. These authors reported an average of 2.6 cuckoo wasps watching throughout the day near each nest. Female *triangulum* were seen to attack cuckoo wasps, but they were unable to inflict injury because of the cuckoo wasps' thick integument and their behavior of rolling into a ball.

In North America, cuckoo wasps of the genera *Hedychrum*, *Hedychridium*, and *Ceratochrysis* have been reported investigating *Philanthus* nests, but no data are available on their life histories. Although some Chrysididae are true parasitoids (Bohart and Kimsey, 1982), the few data available suggest that the association with *Philanthus* (at least in the case of *Hedychrum*) may be that of cleptoparasitism.

By all odds the most abundant natural enemies of beewolves are flies of the family Sarcophagidae, subfamily Miltogrammínae. These flies deposit small maggots on the prey. The maggots consume the host egg and develop on the provisions in the cell; when mature, they dig a short distance from the cell and form their puparia. There are three major genera, which differ in their mode of attack. None are at all host-specific; they attack a wide variety of ground-nesting wasps (Evans, 1970).

Senotainia. The species of *Senotainia* are satellite flies, which follow prey-laden females to their nests (Fig. 2-11) and larviposit on the prey

Major Features of the Biology of Beewolves

Figure 2-11. A female *Philanthus zebratus* approaching her nest with prey. A satellite fly (*Senotainia trilineata*) is visible immediately behind her, an open accessory burrow to the right. (From H. E. Evans and M. J. West-Eberhard, *The Wasps.* Copyright © 1970 by the University of Michigan. Reprinted by permission of The University of Michigan Press.)

as the female pauses at the nest entrance; occasionally they follow the female a short distance inside the burrow. Most North American species of *Philanthus*, perhaps all, suffer mortality from these ubiquitous flies. Often the flies perch near nest entrances, awaiting the return of the female (Fig. 2-12).

Figure 2-12. A satellite fly (*Senotainia* sp.) perched on a nail that has been used to mark a *Philanthus* nest. (From Evans, 1970, courtesy of Museum of Comparative Zoology, Harvard University.)

Metopia. The species of *Metopia* are hole searchers, which are attracted to and enter open holes; larviposition apparently occurs within the cell. Several North American species are known to be attacked by *M. argyrocephala*; *M. leucocephala* is a well-known cleptoparasite of the European *triangulum*.

Phrosinella. The species of the genus *Phrosinella* are also hole searchers, but they also possess small expansions on the front tarsi that permit them to dig through weak closures.

The differing modes of attack of these various enemies are reflected in differing defensive behaviors on the part of beewolf females. *Phrosinella* flies, like cuckoo wasps, appear to be attracted visually to mounds of soil at nest entrances, and the leveling of the mounds may serve to reduce the success of these insects. Accessory burrows and closed nest entrances may serve to divert hole seekers and the oviposition of bee flies; however, closures may delay entry of wasps and thus permit larviposition by satellite flies. These matters are pursued further in Chapter 10.

Economic Importance

The beewolf of Europe, *P. triangulum*, as well as its African subspecies *diadema*, have often been cited as serious predators on domestic honey bees, which may be taken either on flowers or at hive entrances. Thiem (1932) found great numbers nesting in alkali dumps in the Werra Valley of Germany. Many local colonies of honey bees lost most of their foragers, and beekeeping could not be conducted successfully in the area. He reported 150–250 dead bees per square meter. Simonthomas and Simonthomas (1980, p. 99) commented: "One female may catch up to ten honeybees a day. In Europe a *P. triangulum* population of 3000 in an area is no exception, and it is able to capture 30,000 honeybees daily—the number in an average colony.... In Europe, there are many examples of flourishing bee industries that have been destroyed by *P. triangulum*, which has been a serious pest at different times and at various places."

These authors cite a number of such outbreaks and discuss their own work in the Dakhla oasis of Egypt, where the abundance of beewolves had made beekeeping almost impossible. One method of control was to pay a bounty of one Egyptian pound for each 100 wasps captured.

Despite the capture of 24,000 beewolves in one season (1976–77), they were nearly as abundant the next year. Simonthomas and Simonthomas suggested the importation of the cuckoo wasp *Hedychrum intermedium* from Europe as well as the destruction of breeding areas by irrigation, spraying, or covering of the soil with an impervious substance.

In North America, where the honey bee is not native, at least four species of *Philanthus* have been reported as taking honey bees: *bicinctus*, *sanbornii*, *crabroniformis*, and *crotoniphilus*. In most cases, honey bees appear to constitute only a small portion of the prey; but in central Florida, Kurczewski and Miller (1983) found honey bees to constitute 40% of the prey. Many of these honey bees were captured as they left their hives, which were about 40 m from the *Philanthus* aggregation. Clearly there is a risk in housing honey bees in areas suitable as nesting sites for the larger species of *Philanthus*.

The toll of wild bees taken by various species must sometimes be very high indeed. In the case of the so-called bumblebee-wolf, *P. bicinctus*, Gwynne (1981) reported an average of 4.8 bees per cell in a population of about 200 females. Based on a conservative estimate of eight cells per nest, this population must have destroyed over 7500 bumble bees. Great numbers of Halictidae are used by some of the smaller species. The role of wild bees in pollination is little appreciated, and it is true that many pollinate wild flowers and even weeds. The establishment of *Philanthus* populations near agricultural crops could sometimes prove a problem, however, because they might reduce the population of pollinators. The various wasps used as prey by some species of *Philanthus* are mainly beneficial, preying on such insects as caterpillars, grasshoppers, and aphids. On the balance, beewolves should be regarded as undesirable from a human point of view, but in our opinion the fascination of their ways of life more than compensates for their impact on beneficial wasps, pollinators, and commercial honey production.

Chapter 3
Species of the *zebratus* Group

The *zebratus* group includes six species, whose taxonomy has recently been reviewed by Ferguson (1984). They are the largest of North American *Philanthus* species and differ from other species in having their eyes strongly convergent above, a feature that led some persons in the past to place them in a separate genus (or subgenus), *Oclocletes*. The males have dense hair on their abdominal venters that varies from amber to black. Two of the six species have not been studied: *gloriosus* and *ventralis*. The remaining species are considered below in the following order: *bicinctus*, *basilaris*, *zebratus*, and *sanbornii*.

Philanthus bicinctus (Mickel)

P. bicinctus is the largest species of *Philanthus* in North America and possibly the largest in the world; females average 22 mm long and males 17 mm. It is strikingly dimorphic in color, females having a black abdomen with a basal orange band followed by a yellow band, males a more typical "wasp pattern" of alternating black and yellow bands. As a result of this color dimorphism, the males were originally described as a separate species, *hirticulus*. *P. bicinctus* occurs characteristically in the Rocky Mountain area, from Montana to Utah and southern Colorado, mainly at altitudes of 2000–2800 m. Typical habitat is coarse, sandy soil, often overlain or mixed with gravel and often amid plants such as sagebrush, cacti, yuccas, and grasses. This is a species of decidedly localized distribution; although abundant in certain sites, it has rarely been collected elsewhere. Gwynne (1978) has called it the bumblebee-wolf, an appropriate name because the females prey mainly on these large, social bees.

There are several excellent studies of this species, notably those of Armitage (1965), conducted in Wyoming near the south entrance of Yellowstone National Park, and of Gwynne (1978, 1980, 1981), conducted at Great Sand Dunes National Monument in southern Colorado. Shorter reports alluding to this species are those of Mason (1965), Evans (1970), Gwynne and O'Neill (1980), O'Neill and Evans (1982), O'Neill (1985), and Schmidt et al. (1985). The account by O'Neill and Evans pertains to an aggregation at Chimney Rock, Larimer County, in north central Colorado. This section of Chapter 3 is largely a review of the published information, but it adds a few notes relating to each of the three populations that have been studied.

Occurring as it does at fairly high altitudes, *bicinctus* has but a single generation a year. Armitage found it to be active at Yellowstone from mid-July through late August. Gwynne reported first emergence in late July at Great Sand Dunes, with emergence continuing (in a cage placed over the soil near the center of the aggregation) until mid-August; females continued to provision nests until mid-September. At Chimney Rock, the yearly cycle appears similar.

Members of both sexes visit flowers in the vicinity of the nesting area for nectar. Most frequently visited, in our experience, are snakeweed (*Gutierrezia sarothrae*), grey horsebrush (*Tetradymia canescens*), and bushy eriogonum (*Eriogonum effusum*); Armitage reported *Solidago* as a major source of nectar in Yellowstone. Armitage found the peak of feeding activity to occur at about 1100 hours, following an earlier peak of digging (on the part of the females) and preceding the peak of provisioning activity.

Male Behavior

Armitage (1965) first noted that males commonly perch in open spaces and chase one another as well as other insects that fly through the area. He noted two marked males returning to the same perch on successive days and suggested that the behavior had some of the characteristics of territoriality.

The studies of Gwynne (1978, 1980) at Great Sand Dunes, which extended over several years (1975-1978), confirmed that males are indeed territorial. The aggregation here consisted of well over 100 individuals of each sex each year. In 1976 Gwynne marked approximately 200 males with small plastic number tags (*Opalithplattchen*) attached to the mesoscutum (see Fig. 2-1). He found that the males spent the night in abandoned burrows of halictid bees or dug their own shallow burrows (averaging 8 cm long), which were closed from the inside at night. Several males were seen to reuse their sleeping burrows for two or more nights; in some cases more than one male occupied a burrow. Males ranged as far as 250 m from their burrows but returned to them at night. In the morning, 0830–0930, males began to appear at burrow entrances, sometimes leaving briefly and reentering. By 1000 most males had dispersed.

Although males began to emerge about 6–7 days before the first females, they did not establish territories until the females had begun nesting. In the midmorning hours (0930–1100) there was intense competition for spaces clear of vegetation in the nesting area. Eventually the territories became more or less stabilized, each territorial male occupying a perch or a set of close perches on the ground, a stone, or a stick near the center of a fairly bare area 1–3 m in diameter surrounded by tall grasses or forbs. Males often interacted with flying insects farther than this distance from the perch. Gwynne arbitrarily chose a 3-m radius from the perch as the territorial limit for purposes of recording interactions and number of nests per territory. A positive correlation was found between the number of nests within a 3-m radius of a territorial perch and both the number of days the territory was occupied and the average number of hours it was occupied per day.

Gwynne (1980) presented maps showing the position of nests and territorial perches at the Great Sand Dunes site over a 3-year period and over the course of one season. A close correspondence between territory and nest location is apparent in each case. The number of nests per territory varied from 3 to 20, the number of days of territory occupation from 1 to 17. Presumably males respond to the prominent mounds of soil made by females at their nest entrances. Gwynne performed some preliminary experiments to determine whether males would respond to artificial nest mounds made at sites near but apart from the nesting area. On four of six occasions, males established territories at such sites. When the mounds were removed, territorial males did not reappear.

As a result of competition for the limited number of potential sites in the nesting area, many males were excluded from territories. (See data from Gwynne, 1980, in Table 3-1.) However, males also set up territories in bare places in adjacent fields where an abundance of flowering

Table 3-1. Numbers of male Philanthus bicinctus marked, territories used by males, and nests initiated by females over a 3-year period[1]

Year	No. males marked	Total territories used	Max. no. territories used on a single day	Total female nests
1976	187	43	24	182
1977	128	54	34	166
1978	310	79	43	335

1. Data from Gwynne (1980).

plants provided nectar and hunting sites for females. Gwynne termed these "snakeweed territories," snakeweed being the most abundant flowering plant in these areas. Males did not, however, actually defend specific plants in the manner of some male bees (Eickwort and Ginsberg, 1980). There were scattered nests of females in these areas, as noted by Gwynne and as confirmed by us in these same areas in 1980 and 1981, so it is difficult to ascertain what attracted males there—foraging females, nesting females, or both. The behavior of males on these territories was identical to that of males in the nesting area (see below).

Within the nesting area, there was a significant positive correlation between male size and the number of nests within 3 m of the center of the territory. Thus, larger males did succeed in occupying territories containing larger numbers of active females. This finding corresponds with direct evidence from other species of *Philanthus* showing that larger males almost invariably win aggressive encounters on territories (as described in this chapter under the species *basilaris*). Gwynne found no significant difference between the size of territorial males in the nesting area and the snakeweed area, however. Males on snakeweed territories are apparently not males that have been excluded from the nesting area territories because of their size (although we cannot exclude the possibility that there is some unidentified correlate of resource-holding potential in male *bicinctus*). In addition, there is no direct evidence that males moved to the snakeweed area after competing unsuccessfully in the nesting area. Of the 77 marked males that, in 1976, never occupied a territory in the nesting area, 40 were observed on snakeweed territories, although some of these males were seen to use sleeping burrows in the nesting area. Ten other males noted on snakeweed territories were seen at some time on nesting area territories. Thus, at this point, it cannot be determined if males adopt snakeweed territories after losing in competition for sites in the nesting area or if males in the two areas simply exploit alternative sources of receptive females.

At times, individual territories are occupied by a succession of males over the course of the day. The following information from our field notes (made at Great Sand Dunes in 1976) illustrate this behavior: On 15 August, Green 5 occupied territory 31 until 1102. At 1109, Yellow 23 appeared and usurped the territory; Green 5 then established a perch about 1 m away, leaving at 1130. Yellow 23 remained on territory 31 until 1138 but at 1148 had been replaced by White 28, who remained there until 1255.

Gwynne (1978) has described in detail the behavior of territorial males. Such males fly from their perches at intervals to intercept passing insects or to scent-mark grass blades. Gwynne recorded 751 interactions with other insects; a majority were with conspecific males, although numerous interactions involved conspecific females and a variety of wasps, flies, and other intruders. Observed behaviors were categorized in four levels of increasing intensity: (1) approach, (2) butt, (3) brief grapple in midair, and (4) grapple in midair followed by fall to the ground. These categories were assigned values of 1, 2, 3, and 4, respectively. Using these figures, Gwynne calculated a mean response ("response index") for each type of insect encountered. Of the 22 items he listed, the few in Table 3-2 are representative.

In brief, if the intruder was a male, the average response lay between butt and grapple, while if it was a female, the average lay between grapple and grapple to the ground; but grossly unrelated intruders

Table 3-2. Responses of territorial male Philanthus bicinctus to intruders[1]

Intruders	Response index	No. records
Philanthus bicinctus females	3.39	131
Philanthus bicinctus males	2.32	226
Philanthus basilaris (either sex)	1.65	9
Cerceris wasps	1.20	5
Flies	1.00	205
Butterflies	1.00	32
Birds	1.00	7

1. Data from Gwynne (1978).

were merely approached. Gwynne concluded that insects similar in size and color to conspecifics were likely to elicit more intense responses.

Grapples to the ground differed depending upon whether the grappled insect was a male or a female. Grappling males tended to roll over, venter to venter, biting each other. When a male grappled with a female, the male grasped the back of the female's thorax and curved his abdomen in an apparent attempt to initiate copulation.

Copulations were observed only occasionally and were apparently always initiated at the margins of territories. One female flew up from her nest after 45 minutes of digging and mated with a male whose territory was 2 m away. Two other instances were observed in which a female flew up from her nest and mated with a nearby territorial male. Males that had copulated reoccupied territories promptly and were apparently ready to mate again. Although males initially mounted females dorsally, the pair soon assumed an end-to-end position (Fig. 2-5). Complete copulations lasted as long as 6.5 minutes. Females apparently mated soon after beginning to nest.

Scent marking is similar to that of other species but is impressive because of the size of males and the fact that their weight is sufficient to bend the tall grass stems considerably. At Great Sand Dunes, most of the scent marking was on needle-and-thread grass, *Stipa comata*. As described by Gwynne (1978, p. 97), "The body of the male is held in an inverted V position with the distal part of the abdomen and the head touching the stem. Usually he moves down the whole length and often turns around near the base of the plant to move back up the stem." Frequency of scent marking tended to decrease over the course of the day. Gwynne provided a scanning electron micrograph of the male head, showing the clypeal brushes, and also a sketch of a dissection of a male head, showing the enlarged mandibular glands. Schmidt et al. (1985), working with samples from this same site, confirmed that volatiles from the males' mandibular glands were applied to grass stems.

At Great Sand Dunes, *P. basilaris* males tend to establish territories in close proximity to those of *bicinctus*. O'Neill (1983b) reported 11 territorial perches occupied by *basilaris* males on some days and by *bicinctus* on others. Prolonged interactions between the two species were never observed, but *bicinctus* males occasionally chased and made contact with *basilaris* males. As noted in Chapter 2, these two species appear to apply a different blend of pheromones during scent-marking episodes.

Since the publication of Armitage's and Gwynne's papers, we have visited Armitage's site at Yellowstone National Park and have also

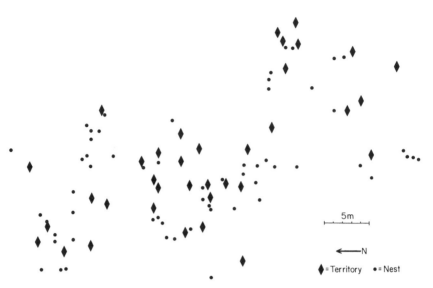

Figure 3-1. Map of a 30-by-50-m section at the east end of *Philanthus bicinctus* nesting area near the South Gate of Yellowstone National Park, Wyo., 5 August 1986. All nests and territories shown were active this day.

studied a small population of *bicinctus* at Chimney Rock, Larimer County, Colorado. In both cases we found nests to be moderately dispersed and males to scent-mark territories in places of high nest concentration, as at Great Sand Dunes. Interactions between males included butting and grappling. Two copulations were observed in Yellowstone in 1986, one lasting 12.2 minutes, after which the male returned to his territory. Distribution of nests and territories at the Yellowstone site in 1986 is shown in Figure 3-1.

Gwynne and O'Neill (1980) showed that male *bicinctus* are much more often preyed upon by asilid flies than are the females. Territorial males may approach asilids and fall prey to them, and asilid flies may actually gather at territorial areas. The major predator at Great Sand Dunes was *Proctacanthus micans* (see Fig. 2-10). In addition, at this same site in 1979 we found a male *bicinctus* as prey in a nest of a female *basilaris*.

Nesting Behavior

Females nest in diffuse aggregations that persist at the same site for many years. Armitage's site at Yellowstone was discovered in 1957 and was still active when we visited the area 24 and 30 years later. Nests are dug in flat or gently sloping soil, often close to the bases of plants. Most are spaced 0.4–1.0 m apart, but we have seen some with entrances only 8–10 cm apart as well as some isolated nests apart from the main nest area. Frequently females continue to occupy the same nest for the entire season, lengthening the burrow and adding cells. Gwynne records a maximum nest duration of 36 days. We have often seen females starting nests later in the season, however; and we assume that some make a second nest if they encounter impediments that make expansion of the burrow difficult.

Most digging occurs in the morning hours, 0830–1300 but especially 0900–1100. Gwynne (1981, p. 132) described digging as follows: "A female starting a nest first chews at the soil surface with her mandibles and as soon as a small depression is made she starts to dig with her front legs. Digging is rapid and after 3–6 minutes the burrow is ca. 2 cm long. The female then uses her wide head to push plugs of soil to the surface, occasionally leaving the burrow to clear soil away from the entrance with her legs. Small stones are pulled from the burrow with the mandibles."

Construction of the oblique burrow requires 3–4 days of digging. Afterward, some digging is done almost daily as the burrow is lengthened and new cells added. Mounds at nest entrances become quite large. They are roughly circular or fan-shaped, measuring 15–21 cm in diameter and 1.5–5.0 cm deep in the center. Each time a digging female appears at the entrance she backs quickly across the mound, then returns slowly to the entrance, digging with her forelegs. The result is a pattern of troughs radiating from the entrance.

Accessory burrows may or may not be constructed at the time of the first nest closure. On 11 August 1975 we counted 163 nests at Great Sand Dunes, of which 14 had one accessory burrow and 5 had two. On 12 August 1976 a census of 127 nests showed that 17 had one accessory burrow, 3 had two, and 2 had three. The overall incidence of accessory burrows was only 14%. These burrows were left open, whereas the true burrow was closed. Most were constructed at an angle of 70–120° to the true burrow and with the opening 2–5 cm from the burrow. They varied in depth from 0.5 to 4.0 cm.

Gwynne reported some nest switching during the first few days of nest construction. Of 22 females marked during 1978, 9 changed to another nest site; however, only 1 of the 9 moved to another nest after spending more than 2 days digging a nest. Several attempts at takeovers of active nests were observed. These resulted in grappling at the nest entrance, the wasps biting each other with their mandibles. Most examples of nest changing involved a female taking over a nest that had been abandoned by another female.

To close the nest, the female emerges and scrapes soil behind her into the hole. At the initial closure, females typically reenter and come out again, making a fresh closure. A slow, ascending orientation flight then takes place. It consists of irregular spirals, at first only about 0.5 m high but gradually rising to a height of 8 m or more and increasing in diameter from 2–3 m to 5–15 m. The entire flight may take 30 seconds, after which the female flies quickly away from the nesting area. Orientation flights are most often seen in the late morning hours, following a bout of digging.

Nests are proclinate and progressive, with a serial arrangement of cells (Fig. 3-2). The oblique burrow measures from 50 to 95 cm long. There may be several lateral bends, as well as undulations, which render the burrow difficult for the researcher to trace. The burrow's diameter is about 1.5 cm at the entrance, narrowing to about 1 cm over most of its length. Cells are built at the ends of short cell burrows, 4–10 cm long, which are constructed on each side, sloping downward from the level of the extended burrow. Distance between neighboring cells varies from 5 to 22 cm. Cells are large, from 20 to 25 mm wide and 25 to 35 mm long; often there is considerable space above the prey in fully provisioned cells. Side burrows are closely packed with soil after cells have been provisioned.

The maximum number of cells found in a nest at Great Sand Dunes was 11, but it is probable that more are sometimes constructed. Cells at this site varied in depth from the surface from 28 to 43 cm (mean = 35.1, $N = 21$); in distance from the nest entrance they varied from 50 to 127 cm. The one nest reported by Armitage (1965) from Yellowstone had a total burrow length of 90 cm; the four cells were about 45 cm in depth from the surface of the soil directly above. Nests at Chimney Rock, Colorado, were all dug on a gentle slope on the side of an arroyo. All seven nests excavated were rather crooked, and all had one or more

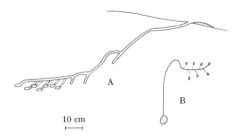

Figure 3-2. Nest of Philanthus bicinctus, Great Sand Dunes National Monument, Colo. (A) Lateral view. (B) Plan view (drawn to a smaller scale than that indicated for A).

places where the burrow passed upward abruptly and then sloped downward again. Burrows varied in length from 62 to 92 cm, with cells constructed at vertical depths of from 33 to 50 cm (mean = 39, $N = 14$).

Provisioning the Nest. Most provisioning occurs in the afternoon, sometimes as late as 1700 hours. It proceeds at a slow pace. Armitage found that the number of bees carried per day varied from one to five. Prey is brought in at a height comparable to the height of the orientation flight. Once in the area, prey-laden females descend obliquely or nearly vertically to their nests.

Once at the entrance, the wasps scrape it open with several strokes of their forelegs. At times they enter directly, with the prey slung beneath them, but more often the prey is released and then drawn in from the inside, often by its antennae. Prey is stored near the bottom of the burrow, frequently in a row, before being transferred to a newly constructed cell. Bees are often moving their legs and antennae as they arrive at the nest. On one occasion one of us was stung by a bee picked up at a nest entrance.

Armitage watched females on clover and goldenrod flying from flower to flower and occasionally attacking bees that sat on flowers or hovered about them. He described prey capture as follows:

> A bee was grasped by a wasp, turned ventral side up and held in the wasp's legs so that the tip of the bee's abdomen could not reach the wasp. The wasp curved its abdomen forward and stung the bee in the thorax. Capture of a bee usually resulted in the bee and wasp tumbling to a lower area of the plant. The wasp, dragging the bee with her hind legs, crawled to the top of the plant. The bee was then held, ventral side up, head anterior, snugly against the wasp's thorax with the middle legs. The wasp rose almost straight up to 30 or more feet above the ground and flew in the direction of the nesting area. (Armitage, 1965, p. 93)

Of 154 wasp-bee interactions observed by Armitage, 23% resulted in capture while 47% resulted in failure after contact was made. In the remaining 30% the wasp avoided the bee at the last moment, suggesting that selection of prey occurs upon close approach, possibly through chemical cues. Two species of syrphid flies that are excellent mimics of bumble bees were consistently avoided by the wasps.

Armitage found eight species of bumble bees to be used as prey in Yellowstone. He collected bees by sweeping vegetation in appropriate places and compared sweep samples with prey samples to determine if the wasps were selecting only certain kinds of bees among those available. The sweep samples differed significantly from the prey samples, males being much more frequently represented in prey samples than might be expected. His samples included about 200 bees taken sweeping and about an equal number of bees taken as prey. Females were without exception workers.

Mason (1965) analyzed Armitage's data and pointed out that in the case of the two larger species of bumble bees there was no difference between field and prey samples, but in the case of smaller species the wasps preferred males, probably because they were larger on average than the workers. No bee with a head width of under 3.06 mm was taken by a wasp, and few with a head width of under 3.15 were taken. He concluded that females selected bees primarily on the basis of size rather than sex.

At Great Sand Dunes, Gwynne (1981) recorded 10 species of bumble bees as prey, including all but 2 of the species found by Armitage in Yellowstone. In contrast to Armitage's findings, Gwynne's records showed that females (workers) made up 85% of prey samples.

Occasional use of Hymenoptera other than bumble bees was reported both at Yellowstone and at Great Sand Dunes (Table 3-3). At Great Sand Dunes during 1976, all prey recorded consisted of bumble bees except for two female *Podalonia* wasps. In 1978, however, when bumble bee populations in the Rocky Mountain region appeared unusually low, only 39% of the prey consisted of bumble bees. The remainder consisted of *Podalonia* wasps, nine species of solitary bees, two worker honey bees, and a female *Psithyrus* (a social parasite of *Bombus*). Although the wasps are obviously able to switch to other prey of suitable size when unable to find bumble bees, Gwynne found that only about half as many items of prey were brought in per day in 1978 as in 1976.

O'Neill and Evans (1982) reported only bumble bees as prey at Chimney Rock, in northern Colorado. Here the majority (86%) were males. Of 14 cells excavated at Chimney Rock, 2 contained four bumble bees, 8 contained five, and 4 contained six. Gwynne reported from two to eight bees per cell at Great Sand Dunes (mean = 4.8, $N = 28$). Three of the cells in the nest reported from Yellowstone by Armitage contained five bees each.

Natural Enemies

As noted above, robberflies, *Proctacanthus micans*, frequently preyed on territorial males at Great Sand Dunes (seven actual records of this behavior were made). There is also one record of a female *Philanthus*

Table 3-3. Prey records for Philanthus bicinctus[1]

Species	Yellowstone Park, Wyo.	Great Sand Dunes, Colo.	Chimney Rock, Colo.
SPHECIDAE			
Podalonia communis		18 ♀	
ANDRENIDAE			
Andrena colletina		1 ♀	
MEGACHILIDAE			
Hoplitis albifrons		1 ♀	
Megachile comata		1 ♀	
Megachile dentitarsus		1 ♂	
Megachile nevadensis		5 ♀ 1 ♂	
Megachile perihirta		3 ♀	
Megachile pugnata		2 ♀	
Megachile sp.	1		
ANTHOPHORIDAE			
Anthophora montana		1 ♂	
Anthophora sp.	1		
Melissodes semilupina		2 ♂	
Melissodes sp.	1		
APIDAE			
Apis mellifera		2 ♀	
Bombus appositus		1 ♀	
Bombus bifarius	47 ♀ 25 ♂	25 ♀ 2 ♂	4 ♀
Bombus centralis	4 ♀	8 ♀	1 ♀ 2 ♂
Bombus fervidus	1 ♂	12 ♀ 2 ♂	
Bombus flavifrons	12 ♂	6 ♀	
Bombus huntii		2 ♀ 6 ♂	7 ♀ 75 ♂
Bombus melanopygus	9 ♀ 9 ♂		
Bombus mixtus	4 ♀ 8 ♂	1 ♀	
Bombus morrisoni		1 ♀	
Bombus occidentalis	73 ♀	3 ♀ 1 ♂	
Bombus pennsylvanicus		1 ♀	
Bombus rufocinctus	4 ♀ 8 ♂		
Psithyrus insularis		1 ♀	
TOTALS 27 species	141 ♀ 63 ♂ (+ 3)	94 ♀ 16 ♂	12 ♀ 77 ♂

1. All female Apidae (other than Psithyrus) are workers, not reproductive females. Non-Apidae from Yellowstone were not identified as to sex.

basilaris preying upon a male *bicinctus* (Gwynne and O'Neill, 1980; *basilaris* was at that time misidentified as *zebratus*). Gwynne has observed occasional predation on both male and female *bicinctus* by spiders. He has a record of a *Geolycosa* preying upon a female as well as a *Steatoda grossa* preying on a male.

Cleptoparasitic flies of the subfamily Miltogramminae appear to be the major pests of nesting females. At Yellowstone, we noted one such fly following a prey-laden female as she descended from high in the air. The fly proved to be *Metopia argyrocephala*, a hole searcher that sometimes follows females to their nests. Gwynne (1981) reported that prey-laden females at Great Sand Dunes were frequently followed to their nests by satellite flies, *Senotainia trilineata*, which often landed on the prey before the wasp entered the nest. Three of 40 cells he examined (7.5%) contained maggots.

Gwynne reported that prey-laden females followed by satellite flies often walked rapidly around the nest mound with the prey. Another defense against parasitism may have been the removal of maggot-infested prey from the cells. It was not uncommon to see discarded bumble bees, sometimes as many as five, lying outside nest entrances (Fig. 3-3). Although these bees were not examined for parasites, Simonthomas and Simonthomas (1972) found that a third of the bees similarly removed from the nests of *P. triangulum* had been parasitized by flies.

A number of Hymenoptera present in the *bicinctus* nesting area at Great Sand Dunes are known parasites of ground-nesting wasps and bees and may have been attacking *bicinctus*, although there are no unequivocal records of such activity. Cuckoo wasps, *Hedychrum violaceum*, were occasionally seen at nest entrances. Two species of Mutillidae were common in the area, *Dasymutilla ursula* and *D. vestita*.

Figure 3-3. Mound and nest entrance of *Philanthus bicinctus*, showing bumble bees that have been dragged from the nest and discarded.

Females of the latter species were often seen exploring holes, but as the species has been recorded as a parasite of anthophorid and megachilid bees, its association with *Philanthus* nests may have been fortuitous.

Philanthus basilaris Cresson

Like *bicinctus*, *basilaris* is a species characteristic of the Rocky Mountain area. In fact, at our two major study areas, the two species occurred together, although showing only slight overlap of nesting and territorial sites. These two study areas were Great Sand Dunes National Monument, Alamosa County, Colorado, and Chimney Rock, Larimer County, Colorado. Both are at an elevation of about 2500 m. *P. basilaris* also ranges onto the high plains east of the Rockies, and we have studied it briefly on the Pawnee National Grasslands, Weld County, Colorado, at an elevation of about 1500 m. Typical habitat is open country with few trees, carpeted with grasses, cacti, yuccas, and various flowering plants, where small spaces among vegetation offer soil that is moderately friable. At Great Sand Dunes, nests were dug in coarse, sandy soil well away from the major dune mass. At Chimney Rock, nests were dug in coarse, reddish sand resulting from the degradation of nearby sandstone cliffs.

P. basilaris is a relatively large species of the genus, although smaller than *bicinctus*. Both sexes have the abdomen banded with black and yellow; sexual dimorphism in color is much less pronounced than in *bicinctus*. O'Neill (1983b) has published a detailed report on the behavior of males at the Great Sand Dunes site. O'Neill and Evans (1981) also reported on predation on conspecific males by females at this site, and O'Neill and Evans (1982) discussed aspects of predatory behavior at the Chimney Rock site. Earlier reports from Colorado identified this species as *zebratus*, but that is now regarded as a separate species (Ferguson, 1984); *basilaris* is closely similar to *zebratus* structurally but differs in major aspects of behavior. Thus, Gwynne's comments on "*zebratus*" in his papers on *bicinctus* (1978, 1980), Gwynne and O'Neill's paper (1980) on sex-biased predation, and the Colorado (but not Wyoming) populations mentioned by Evans and O'Neill (1978) with respect to alternative mating strategies apply properly to *basilaris*. The present report attempts to synthesize published information and to fill in details not previously reported.

In Colorado, *basilaris* is characteristic of late summer and clearly has but a single generation a year. A major nectar source is bushy Eriogonum (*Eriogonum effusum*). At Great Sand Dunes we have also seen adults taking nectar from grey horsebrush (*Tetradymia canescens*), and both at that site and at Chimney Rock from snakeweed (*Gutierrezia sarothrae*).

Male Behavior

Males spend the night and periods of inclement weather in short "sleeping burrows" dug in the same general area in which the females nest. In the mid or late morning hours they emerge and move about, taking nectar and apparently inspecting potential territory sites. O'Neill (1983b) followed several males from the time they were first seen to the time they settled and remained on a territory. As he described it, "One male moved four times in three minutes between three different perches before remaining on one. A second male moved ten times in seven minutes between five perches, while a third moved eight times in three minutes between five perches. Upon choosing a perch, the males immediately begin scent-marking and defending it from conspecific males" (p. 308).

Figure 3-4. A male *Philanthus basilaris* on his territorial perch on a prickly pear cactus, Great Sand Dunes National Monument, Colo.

Territorial perches include patches of bare soil, clumps of *Opuntia* cacti, or the mounds made by harvester ants, where there are surrounding grasses or other vegetation 30–80 cm tall (Fig. 3–4). Territories are 1.0–1.5 m in diameter and are defended vigorously, particularly within the first hour after they have been established (Fig. 3–5). The rate of scent marking is considerably lower than in many species of *Philanthus* and declines rapidly after the male has occupied the territory for some time. O'Neill (1983b) recorded a mean of 10 scent marks during the first 30 minutes after initiation of a territory; at 60 and 90 minutes the mean was about 4, and at 120 minutes about 2. About one time in five, the wasps marked two stems in a single bout, usually closely adjacent stems.

The following record describes a full day's occupation of one territory by a marked male at Great Sand Dunes. This male remained on the territory for 2 hours and 26 minutes (minus a break of 6 minutes, presumably to obtain nectar). During the first hour he scent-marked 16 times in 14 bouts, during the second hour 4 times in 4 bouts, and during the last 26 minutes only once. There were 34 recorded interactions with other male *basilaris* during the course of the day as well as 23 interactions with other insects (bees, flies, and one beetle). There were also many swift flights from the perch not in obvious response to any stimulus; but many may have been, in fact, responses to passing insects we did not see.

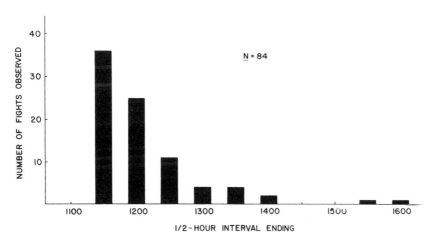

Figure 3-5. Frequency distribution of territorial contests of male *Philanthus basilaris* observed in each 30-minute interval (ending at time indicated). Data combined from 5 days in 1979 include only interactions that involved physical contact between males. (From O'Neill, 1983b, *Behaviour*, courtesy of E. J. Brill, Leiden.)

The mean net duration of occupation for males that spent brief periods away from their perches (probably at flowers) was 121 minutes ($N = 17$); the equivalent value for males that did not leave their territories was only 78.6 minutes ($N = 14$); the mean duration of occupation for males that had their territories usurped was only 14 minutes ($N = 4$) (O'Neill, 1983b).

Interactions with conspecific males are unusually intense and frequently involve butting, grappling, and biting. Butting occurs when a male strikes a flying or perched male and may produce a clicking noise audible some distance away. Grappling consists of wrestling on the ground with attempts to grasp one another with the mandibles. Contests may occur between territorial residents and intruders or between males on adjacent territories. The latter often occur when a male making a pursuit flight enters a neighboring territory, and they may end with one male abandoning his territory.

In 1979, O'Neill (1983b) recorded 46 aggressive interactions between territorial residents and intruders of known size. All of the interactions included physical contact, and at least 77% included grappling. The larger of the two contestants won all of the interactions that involved males of unequal size ($N = 43$); intruders won 37% of the 46 interactions, thus assuming sole possession of the territory. In 7 interactions between residents of known size on adjacent territories, the larger male was always the victor. Thus, in 50 contests between males of different sizes, the larger male won 100% of the encounters, regardless of his status as resident or intruder.

Small males (with head widths less than 3.2 mm) were not commonly found on territories. Although they made up 34% of the population at Great Sand Dunes, they occupied only 11% of the territories. Conversely, males with head widths greater than 3.3 mm occupied 46% of the territories, although they represented only 21% of the population. The mean head width of all territorial males (3.33 mm, $N = 747$) proved significantly greater than that of nonterritorial males (3.02, $N = 71$) (t-test, $t = 11.55$; $P \ll 0.001$) (Fig. 3-6).

O'Neill further documented these results by removing males from territories and recording the sizes of the new occupants. In 1979 and 1980, the removed resident was replaced by another male within 30 minutes on 30 occasions; on 15 of these occasions, the second male arrived within 5 minutes. In all cases in which the replacing male was different in size from the resident, it was smaller (Fig. 3-7). Nonterritorial males, termed *floaters*, moved through the area and assumed

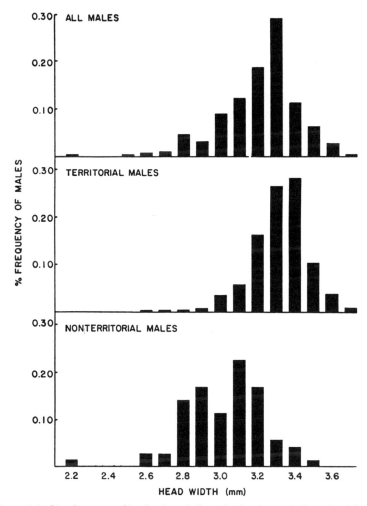

Figure 3-6. Size-frequency distribution of all marked males and all territorial males of *Philanthus basilaris* in 1979 and all nonterritorial males in 1979 and 1980. (From O'Neill, 1983b, *Behaviour*, courtesy of E. J. Brill, Leiden.)

residency if the owner left or was removed experimentally (O'Neill, 1983b). They also challenged residents for ownership of territories.

At Great Sand Dunes, territories of *basilaris* and *bicinctus* were sometimes intermingled, and 11 territories were found to be occupied by a male of one species on one day and a male of the other on another day. *P. bicinctus* males occasionally chased and made contact with *basilaris* males, but no prolonged interactions were observed. At the

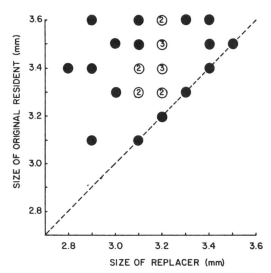

Figure 3-7. Relative sizes of original and replacing males of *Philanthus basilaris*, based on 30 removal experiments. Numbers in circles indicate multiple records. Dashed line indicates equal-sized residents and replacers. (From O'Neill, 1983b, *Behaviour*, courtesy of E. J. Brill, Leiden.)

Chimney Rock site, *basilaris* males were seen to chase the much smaller males of *barbiger* from their territories.

Many of the details of territorial behavior were confirmed at the Pawnee and Chimney Rock sites. Males were uncommon at the Pawnee, and few male-male interactions were observed. One male that was watched for 1 hour (1043–1143) marked 15 plants within a 1.5-m radius of his perch. Scent marking occurred 17 times within the hour, and there were eight pursuit flights after flies and wasps. At Chimney Rock, males were more plentiful, and frequent male-male interactions were observed.

We failed to find nests of the females at the Pawnee site, so we cannot judge the relationship of territories to nests. At Chimney Rock, nests were scattered along the upper slope of a gully and on flat soil close to the gully (O'Neill and Evans, 1982). In 1981, males were grouped at two territorial areas, both north of the gully, about 40 m apart, and 12–30 m from the nearest known nests. In 1984, males were abundant at one of these same sites and also along a little-used dirt road south of the gully, at least 50 m from the nearest known nests. In each area, 10–30 males occupied roughly adjacent territories.

The tendency of males to group their territories in leks was studied in detail at Great Sand Dunes (O'Neill, 1983b). This mating system is uniquely suited to the very diffuse nests of this species. At Great Sand

Species of the *zebratus* group 51

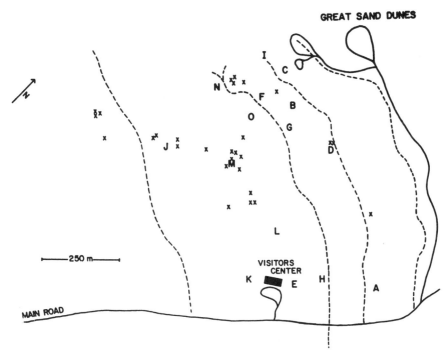

Figure 3-8. Map of *Philanthus basilaris* study site at Great Sand Dunes National Monument, Colo. Solid lines indicate roads and dashed lines gulleys. Leks are lettered A–O; x indicates nests found in 1978 and 1980. (From O'Neill, 1983b, *Behaviour*, courtesy of E. J. Brill, Leiden.)

Dunes, females nested over a wide expanse of the sandy fields adjacent to the dunes (Fig. 3-8). No lone territorial males were found; rather, all territories were clustered in places offering no obvious resources other than the males themselves—that is, no nesting sites or unusual concentrations of flowers serving as sources of nectar or prey.

In 1980, we found 14 aggregations of territories at Great Sand Dunes. They were widely spaced, between 70 and 200 m from their nearest neighbors. Distances between territories within aggregations averaged only about 2 m. Of the 14 leks, 10 usually had from 5 to 10 males, 3 contained between 10 and 20 males, and 1 contained more than 20 on most days. The largest aggregation contained a maximum of 84 territories, with from 5 to 47 active on any one day (mean = 28.1, N = 14 days). Several of these territorial sites were reoccupied by successive yearly generations of males, one of them for at least 5 years (1977–1981). Observations in 1981 (not reported in O'Neill, 1983b) showed 6 of the 14 lekking sites of 1980 to be occupied.

Each aggregation tended to develop through the season from a small core of territories, eventually assuming an elongate, elliptical shape, with the main axis of the ellipse parallel to the prevailing wind direction, which was fairly constant at Great Sand Dunes. Most new territories were established downwind of the older territories as the season progressed. Possibly males find these sites through detection of pheromone from downwind, then set up their territories downwind of existing territories, thus taking advantage of pheromone from these territories. Females may also be expected to approach from these territories.

Patterns of territory use demonstrated that territories were not occupied equal numbers of days. Table 3-4 illustrates variations in patterns of occupancy for two aggregations studied in 1979.

One might expect that larger males would tend to occupy territories of highest quality within the leks and that these territories would be on the downwind side or possibly in the center (as reported for vertebrate leks). We were unable to demonstrate a difference in mean size when we compared males from upwind, downwind, and central parts of individual leks, however. Furthermore, we found no correlation between frequency of occupation and either distance to the nearest flowering *Eriogonum* plant or distance to the nearest territory. There was, however, a low but significant correlation between the number of days a territory was occupied and the size of males on the territory for both of the aggregations referred to in Table 3-4 (O'Neill, 1983b).

Mating Behavior. In three years of study at Great Sand Dunes, we observed only five matings. All occurred at leks between 1152 and 1350

Table 3-4. Number of days that individual territories were occupied by male Philanthus basilaris in two different aggregations[1]

Aggregation C (censused 22 days)		Aggregation F (censused 14 days)	
No. days occupied	Frequency	No. days occupied	Frequency
1-3	12	1-2	32
4-6	6	3-4	16
7-9	5	5-6	13
10-12	2	7-8	13
13-15	1	9-10	6
16-17	1	11-12	4
Total	27	Total	84

1. Data from O'Neill (1983b).

hours. In two cases, we saw the females before coupling was initiated; in both cases they flew upwind into the territorial area at heights of 2–3 m.

> In one case, a male flew up from his perch, grappled the female, coupled with her in mid-air, and landed with her on a grass stem two meters from his territory. They remained motionless and coupled for a total duration of 5.5 minutes. In the other mating, two males simultaneously grappled the incoming female in mid-air. After landing on a plant four meters away, both males were coupled with her. Fifteen seconds later, a third male pounced on the trio causing all four to fall to the ground. After further struggle, one male broke free with the female and completed copulation. The entire sequence from first contact with the female to completion of mating lasted 15 minutes. At least four of the five copulations were completed away from the territory the male had been defending. At least twice, the males returned to their territories after mating. (O'Neill, 1983b, pp. 301–302)

Nesting Behavior

As already noted, the nests of females tend to be dispersed widely. All nests at Great Sand Dunes were in flat or nearly flat soil, but some of those at Chimney Rock were in a soil sloped up to 30°. At Great Sand Dunes, we found two nests only 18 cm apart, and others were only 1.5–5.0 m apart; but by far the majority were many meters from other known nests. In all, we were able to find here about 30 nests scattered over an area about 500 by 100 m. About the same number of nests were found at Chimney Rock in an area measuring about 60 by 140 m; some nests were separated by no more than half a meter, but most were more widely spaced than that. Because of their dispersed nature, it was difficult to locate and mark all nests, and we undoubtedly missed many, especially at Great Sand Dunes.

Mounds at nest entrances are not leveled and remain conspicuous unless eroded away. Mounds vary in length from 6 to 16 cm, in width from 5 to 10 cm. Digging females often permit the soil to block the entrance from the inside, then emerge periodically to clear the entrance by scraping soil over the mound, often producing one or more grooves through the mound. More than one day is required to dig the deep burrow through the coarse soil of the substrate.

Accessory burrows are of variable occurrence. At Chimney Rock, only three of eight nests were recorded as having these burrows, which varied in depth from 1 to 3 cm. At Great Sand Dunes, six of seven nests were recorded as having accessory burrows, varying in depth from 1 to

11 cm. Four of the six nests had two accessory burrows, one on each side of the true burrow.

The burrow measures about 6 mm in diameter and penetrates the soil at a 30–50° angle to the horizontal (but may approach 90° to the angle of a slope). Deep in the soil it tends to be nearly horizontal, with gradual or abrupt undulations and frequent lateral turns (Fig. 3-9). The oblique burrow varies from about 50 to 70 cm in length, but it may eventually be extended to about a meter. Cells vary in distance from the burrow entrance from 50 to 100 cm. Cell depth at Great Sand Dunes varied from 28 to 46 cm (mean = 36.4, N = 21), at Chimney Rock from 20 to 47 cm (mean = 36.4, N = 18).

We found from one to seven cells in the nests we excavated at the two sites. The seven-celled nest had received a final closure, and all its cells contained cocoons. We suspect that some females make more than one nest per season. On 1 September 1978 we checked 12 nests that had been marked 3 weeks earlier, but none were still active, although numerous other nests were active on that date. Cells measured about 15 by 25 mm and were constructed at the ends of rather long cell burrows (10–30 cm) that branched from the main burrow. Nests were proclinate, but we found no regular pattern of cell construction as in *bicinctus*. We found no apparent tendency for newer cells to be farther from the entrance than older cells.

Provisioning the Nest. We have observed females hunting at snakeweed, hovering downwind from blossoms and moving from plant to plant. One predation on a bee was observed at an *Eriogonum* plant in 1980, after the female had approached it from downwind. Females also hunt for males of their own species at lekking sites, at least occasion-

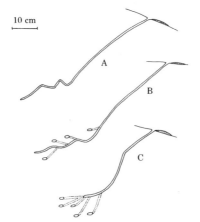

Figure 3-9. Profiles of nests of *Philanthus basilaris*. A is a fresh burrow, without cells. A and C are from Chimney Rock, Larimer Co., Colo.; B is from Great Sand Dunes National Monument, Colo. Cell burrows had been filled and those drawn (dashed lines) are hypothetical.

ally. As reported by O'Neill and Evans (1981), females have been seen to enter these sites and be approached by males, resulting in grappling and the eventual stinging of the male by the female, who then carries the male off to the nest. Two conspecific males were found in a cell of a *basilaris* nest at Great Sand Dunes.

Provisioning females descend to the nest from a considerable height, scrape open the entrance, and enter quickly, holding the prey beneath them. When followed by satellite flies, females may fly away a considerable distance, return a few moments later, and enter.

Prey consists of bees of moderate size but also includes many wasps (Tables 3-5, 3-6). At Great Sand Dunes, 42% of the prey consisted of

Table 3-5. Wasps taken as prey by Philanthus basilaris (all Colorado)

Species	Great Sand Dunes	Chimney Rock
ICHNEUMONIDAE		
Diphyus sp.		1 ♀
Pterocormus sp.		1 ♂
Genus & species?		1 ♀ 2 ♂
TIPHIIDAE		
Myzinum maculatum	1 ♀	
Paratiphia sp.	1 ♀	
VESPIDAE		
Ancistrocerus sp.	1 ♀	
Stenodynerus sp.	2 ♀	2 ♀
FORMICIDAE		
Formica sp.		1 ♂
POMPILIDAE		
Anoplius dreisbachi		1 ♀
SPHECIDAE		
Ammophila azteca		1 ♀
Bicyrtes ventralis	1 ♀	
Cerceris vicina	1 ♂	
Cerceris wyomingensis	2 ♂	
Eucerceris cressoni		1 ♂
Larropsis vegeta	1 ♀	
Microbembex monodonta	27 ♀	
Philanthus basilaris	4 ♂	
Philanthus bicinctus	1 ♂	
Podalonia communis		1 ♀
Podalonia mexicana	1 ♀ 1 ♂	
Podalonia mickeli		3 ♀ 1 ♂
Podalonia sp.		3 ♀
Prionyx canadensis		1 ♂
Prionyx parkeri		1 ♂
Stictiella plana	4 ♀	
TOTALS 25 species	39 ♀ 9 ♂	13 ♀ 8 ♂

Table 3-6. Bees taken as prey by Philanthus basilaris (all Colorado)

Species	Great Sand Dunes	Chimney Rock
COLLETIDAE		
Colletes phaceliae	2 ♀ 2 ♂	1 ♀ 2 ♂
Colletes simulans		3 ♀
Colletes spp.	16 ♀ 23 ♂	9 ♀ 1 ♂
ANDRENIDAE		
Andrena colletina		1 ♀
Andrena surda		3 ♀
Andrena sp.	1 ♀	1 ♀
HALICTIDAE		
Agapostemon angelicus		1 ♂
Agapostemon splendens	2 ♂	
Agapostemon texanus		1 ♀ 15 ♂
Dialictus pictus	1 ♀	
Dialictus succinipennis	2 ♀ 6 ♂	
Dialictus pruinosiformis		1 ♀ 5 ♂
Evylaeus cooleyi	1 ♀	
Evylaeus sp.		1 ♂
Halictus confusus		3 ♀
Lasioglossum sisymbrii	2 ♀	1 ♀ 2 ♂
Lasioglossum trizonatum		5 ♂
Sphecodes spp.	1 ♀	1 ♀ 5 ♂
MEGACHILIDAE		
Heriades sp.		1 ♀
ANTHOPHORIDAE		
Anthophora maculifrons	1 ♀	
Ceratina neomexicana	1 ♀	
Epeolus minimus	2 ♀	
Epeolus spp.		8 ♀ 1 ♂
Melissodes grindeliae		2 ♂
Melissodes microsticta		1 ♂
Melissodes pallidisignata		2 ♂
Melissodes spp.	1 ♂	2 ♀ 6 ♂
Nomada sp.	1 ♀	
APIDAE		
Bombus huntii		2 ♂
TOTALS 29 species	31 ♀ 34 ♂	36 ♀ 51 ♂

wasps, including especially *Microbembex monodonta*, an abundant ground-nesting wasp at that site. One nest contained 21 female *Microbembex* in four cells, six in one cell with one additional item of prey, a bee. Most cells contained a mixture of wasps and bees. At Chimney Rock, only 19% of the prey consisted of wasps. At both sites, more female than male wasps were taken, but more male than female bees. The number of prey per cell at Great Sand Dunes varied from 4 to 14 (mean = 7.8, N = 13), at Chimney Rock from 5 to 11 (mean = 6.7, N = 13).

At Great Sand Dunes, *basilaris* took appreciably smaller prey than *bicinctus* and larger prey than *psyche* and *pulcher*, which nested nearby. At Chimney Rock, too, *basilaris* took smaller prey than *bicinctus*; the only overlap involved two male bumble bees (the major prey of *bicinctus*) found in one *basilaris* nest. At this site there was almost no overlap in prey size with the much smaller species *barbiger* but much overlap with *inversus*, a species similar in size to *basilaris*. However, *inversus* is a specialist on males of the genus *Agapostemon* (Halictidae), which are only occassionally used as prey by *basilaris* females (O'Neill and Evans, 1982).

Natural Enemies

Of 18 cells excavated at Chimney Rock, 2 contained maggots of miltogrammine flies and 3 had had the cell contents destroyed by small ants. *Senotainia trilineata* flies were seen following prey-laden females occasionally. These flies were present at Great Sand Dunes but were not actually seen following females, and none of the 21 cells excavated appeared to have been attacked by miltogrammines or other parasites or predators.

Adult *basilaris* males were, however, attacked by asilid flies at Great Sand Dunes; four such records are given by Gwynne and O'Neill (1980). As mentioned above, female *basilaris* are known to prey on males of their own species (O'Neill and Evans, 1981). Of 16 attempted conspecific predations observed, 6 were successful. The 6 males killed ranged in head width from 2.8 to 3.2 mm, whereas the 10 that escaped ranged from 3.3 to 3.6 mm.

Philanthus zebratus Cresson

P. zebratus is closely related to *basilaris*, and there have, in the past, been difficulties in distinguishing the two species. *P. zebratus* is characteristic of the northern Rocky Mountain area, ranging from Wyoming north to Alberta and west to British Columbia, Oregon, and the Sierras of California. We have studied it at two sites in northwestern Wyoming and have published two papers on the unique male behavior: Evans and O'Neill, 1978 (Wyoming records only; Colorado records now properly apply to *basilaris*) and O'Neill and Evans, 1983a. O'Neill (1985) discussed prey and eggs of females. Evans (1966b, 1970) discussed female nesting biology briefly, under the name *P. zebratus nitens*. It is apparent, however, that *nitens* is no more than a color variant of *zebratus* and

58   The Natural History and Behavior of North American Beewolves

unworthy of taxonomic recognition. Ferguson (1984) presented prey records from museum specimens collected in Alpine County, California. This section of Chapter 3 gives a synthesis of published information, with additional data on nesting behavior.

Our major study site has been Deadman's Bar, 14 km southwest of Moran Post Office in Grand Teton National Park, Wyoming. Our initial studies were made in 1964, and we returned for further studies in 1967, during every summer from 1976 through 1981, and again in 1986; thus this aggregation has been active at this site for at least 23 years. The site is an area of alluvial sand along a dirt road within 100 m of the Snake River. Vegetation is somewhat sparse and consists of grasses, sagebrush, and low flowering plants; partially surrounding the area are groves of aspen, pine, and spruce trees. Brief observations were also made on a smaller aggregation in 1981, this one located some 40 km north of this site, at Huckleberry Hot Springs. The nesting area here was also along a dirt road in sandy soil where there was relatively sparse, low vegetation and adjacent groves of trees.

The aggregation at Deadman's Bar became active each year during the first week of July and remained active, in diminished numbers, into mid-August. The degree of protandry appeared slight relative to *bicinctus*, males appearing only a day or two before the first females. Nests were initiated soon after the first females appeared, and male reproductive behavior was initiated as soon as females began to dig. Major nectar sources for both sexes were species of *Eriogonum* and *Solidago* growing in and near the nesting area.

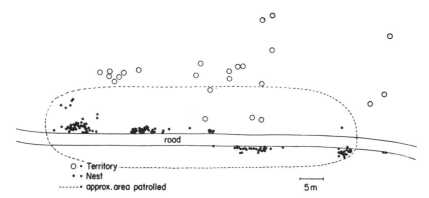

Figure 3-10. Map of study site of *Philanthus zebratus* at Deadman's Bar, Grand Teton National Park, Wyo., showing location of territories, nests, and area patrolled by males in 1980. (From O'Neill and Evans, 1983a, reproduced with permission from Academic Press London on behalf of the Linnean Society, *Biological Journal of the Linnean Society* 20: 175–184, © 1983.)

Species of the zebratus group

The size of the population at Deadman's Bar exceeded 100 individuals of each sex during each year that we studied it. In 1980, eight days after the appearance of the first nest, we marked 178 males and counted 151 nests of females. In 1981 we counted 186 nests on 16 July, but on 23 July only 47 active nests could be found. Nests tended to be clumped at several places along the sides of the road, where entrances were often only a few cm apart (on 13 July 1979, mean = 0.87 m, SD = 0.93) (Fig. 3-10). The entire nesting area measured about 15 by 60 m. The population at Huckleberry Hot Springs was much smaller; we counted only 23 nests, and they were separated by from 0.5 to 45 m (mean = 4.9 m, SD = 3.8).

Male Behavior

Males at the Deadman's Bar site exhibited a dual mating tactic, some participating in a patrolling flight high above the nest area and others establishing territories somewhat apart from the nests. Males spent the night in short burrows along the edge of the road, often among nests of females (Figs. 3-11, 6-6F). These burrows were closed from the inside at

Figure 3-11. Sleeping burrows from which male *Philanthus zebratus* have emerged, found alongside road shown in Figure 3-10.

night, and each had a small mound of soil outside the entrance; they were oblique and measured from 4 to 11 cm long (mean = 8.3, $N = 4$). After they emerged in the morning, between 1000 and 1100 on sunny mornings, males fed at flowers and perched on the ground and on low plants among the nests. They sometimes perched only a few centimeters apart, but they showed no site attachment and no tendency while perching to interact with conspecifics or other insects. Males often remained motionless on plants at this time, in one case as long as 30 minutes.

Between 1113 and 1155, males began to rise in the air periodically. From then until middle or late afternoon there was a canopy of males flying 3–5 m above the nesting site. The flight of each male began with a quick ascent that was vertical or nearly so. After males reached a height of 3–5 m they assumed horizontal flight paths, hovering or flying slowly over a distance of 1–10 m. During this flight they often rose or fell a short distance and turned frequently. There was no evidence that males followed a set flight pattern or patrolled a circuit or territory. They did, however, often change speed rapidly, from a hover to a quick interaction with another insect. Upon completion of a flight, they dropped quickly to the ground, either vertically or obliquely. During 1981, we found the duration of the flights to vary from 3 to 18 seconds (mean = 9.9, $N = 61$); the time spent on the ground between flights varied from 6 to 23 seconds (mean = 12.9, $N = 110$).

We observed flying males against a background of dark green spruce trees, which made it possible to follow individual males on many occasions. Each day numbers increased until noon or early afternoon, then tapered off later in the afternoon (Fig. 3-12). When in flight, males approached or pursued any fast-moving object that passed near them, most commonly conspecific males but sometimes other insects and on one occasion a hummingbird. They readily changed their flight paths to pursue stones that we threw in the air at a suitable height. Individual males moved about within the swarm and usually did not return to the place on the ground from which they had started.

Males that approached one another in the swarm often swirled about one another in tight circles. In other cases one would pursue the other a short distance, occasionally to the ground if the pursued male descended. In none of these cases did we observe obvious butting or grappling. The peak of male flying activity each day spanned the time of most female orientation flights as well as their initial returning flights with prey, both of which occurred at the height of the male patrolling

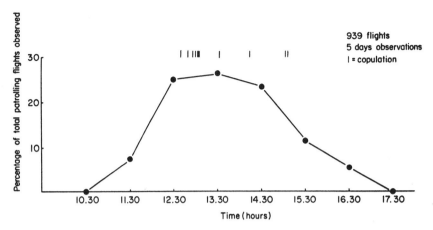

Figure 3-12. Daily schedule of number of patrolling flights of *Philanthus zebratus* observed in 10-minute intervals through 5 days. (From O'Neill and Evans, 1983a, reproduced with permission from Academic Press London on behalf of the Linnean Society, *Biological Journal of the Linnean Society* 20: 175–184, © 1983.)

swarm. Thus we assume that the swarm functions to intercept females when they are most likely to be receptive.

We initially failed to observe territorial behavior among males of the Deadman's Bar populations and assumed that all participated in the high patrolling flights (Evans and O'Neill, 1978). Further study in 1978–1981 revealed that on some days several males had set up territories somewhat apart from the nesting area (O'Neill and Evans, 1983a; Figs. 3-10, 3-13). The number of males territorial on any one day varied from 0 to 16. Territories were in bare places among vegetation and were scent-marked in a manner similar to that used by other species of *Philanthus*.

One male we watched for a full hour (1213–1313) scent-marked 20 times during the first 10 minutes, 9 times during the second 10 minutes, 5 times during the third, but only 6 times in the remaining 30. Males pursued passing insects and interacted vigorously with conspecific males, engaging in butting and grappling. Four of seven interactions observed in 1979 resulted in usurpation of the territory by the intruder. In five instances these interactions involved males of known size, and in all five the larger of the two won the contest. The mean head width of the winners was 3.23 mm (SD = 0.11), that of losers 2.77 (SD = 0.21) (Mann-Whitney test for difference, $P < 0.001$). Clearly, larger males have the advantage in territorial contests, as in other species of *Philanthus*.

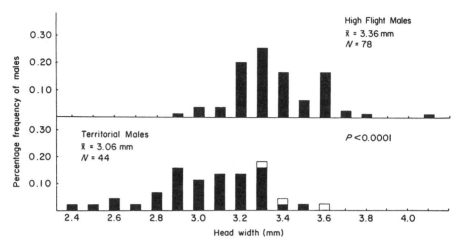

Figure 3-13. Size-frequency distribution of patrolling males and territorial males of *Philanthus zebratus* in Jackson Hole, Wyo., in 1979 and 1980. Hollow portions of bars in lower histogram indicate the size of three males that both held territories and made patrolling flights. (From O'Neill and Evans, 1983a, reproduced with permission from Academic Press London on behalf of the Linnean Society, *Biological Journal of the Linnean Society* 20: 175–184, © 1983.)

Comparison of the size of territorial males with males participating in patrolling flights proved particularly interesting. In 1979 and 1980, 78 males were caught after they had been seen to make patrolling flights. They had a mean head width of 3.36 mm (SD = 0.20; Fig. 3-13). In contrast, 44 males found to be territorial had a mean head width of 3.06 mm (SD = 0.26). The means were significantly different (t-test, $P < 0.001$). Three of these 122 males were seen to patrol at one time and to be territorial at another time. These three males had head widths of 3.3, 3.4, and 3.6 mm and were thus among the largest territorial males, though of average size as patrollers. We concluded that *zebratus* males practice a conditional strategy based on size and that in this dense nesting aggregation patrolling was the primary strategy—that is, the one most likely to result in finding a mate.

Mating Behavior. We observed 11 matings by patrolling males and none by territorial males. The matings occurred between 1238 and 1505, during the height of the patrolling flights. It is true that initiation of matings on territories is less conspicuous, and we did not spend many hours at territories. Pairing of males and females probably occurs primarily at the level of the aerial swarm, but the pair quickly settles to

Figure 3-14. A mating pair of *Philanthus zebratus* on a grass stem beneath the area of patrolling flights. The female is uppermost.

the ground. Nine of the 11 mating pairs were found after they were already at ground level; the other 2 were first observed at heights of 2 and 5 m.

One copulation watched in its entirety lasted 6 minutes, 20 seconds. The pair descended from about 2 m, the male mounted on the female's back, facing forward. They landed on a grass stem 0.3 m above the ground, where the female faced upward and the male turned around and faced downward. In 10 seconds the female took flight to another grass stem, dragging the male behind. Twenty seconds later they flew to another grass stem, 0.7 m tall, and assumed the same posture (Fig. 3-14). Eventually the male pulled away and flew off, the female also flying off after several seconds.

Several male-female interactions not leading to copulation were observed. In three cases males were seen to contact females at 3–4 m height, the pair dropping to the ground, or, in one case, the female being chased the entire length of the nesting area. In another case a male struck a prey-laden female at about 3 m height, deflecting her from her flight path but not causing her to lose her prey.

Male head width was recorded in 7 of the 11 copulations seen; it ranged from 3.2 to 3.6 mm (mean = 3.41 mm, SD = 0.17, $N = 7$). These data include one male recorded mating twice. Although the sample is small, note that all of these individuals were larger than the average for the population.

Observations at a Second Aggregation. As noted above, the aggregation at Huckleberry Hot Springs was smaller and more diffuse than that at Deadman's Bar. We counted a total of 23 nests and 12 territories. Unfortunately we studied this population for only 3 days, over the period 17–23 July 1981. We observed no high, patrolling flights during this time. Eight of the territories were located within or close beside the nesting area. The other four were 50 m or more away from this site, one of them about 100 m away, on a pile of horse droppings in a meadow separated from the nesting area by several trees (Fig. 3-15). We measured the head width of 12 males that held territories and found them to range from 2.8 to 3.4 mm (mean = 3.13 mm, SD = 0.18). This mean was not significantly different from that obtained for all males in the Deadman's Bar population. It seems probable that under the conditions of low nest density prevailing here, patrolling may be less effective. Conversely, under conditions of high nest density, the cost of territorial defense may be prohibitively high.

Behavior of the territorial males at Huckleberry Hot Springs did not

Figure 3-15. A male *Philanthus zebratus* that has established a territorial perch on a pile of horse droppings.

differ notably from that of the territorial males at Deadman's Bar. One male, watched for 10 minutes (1215–1225) on 19 July, scent-marked 16 times in 15 bouts; marking consisted of a short upward movement (ca. 5 cm) followed by a longer movement downward along the stem (15–30 cm). Although this male scent-marked two stems on one bout only once, another male was seen to mark four stems in one bout. The first male interacted with a male from a neighboring territory seven times in 10 minutes, all brief swirling flights except for one grapple to the ground.

Nesting Behavior

As mentioned before, nests at Deadman's Bar tended to be clumped with a mean nearest-neighbor distance of less than 1 m. Nests are conspicuous because of the large mound at the entrance. This mound is not leveled but may have one or more grooves passing across it from the nest entrance. Mounds measure 10–16 cm long by 9–15 cm wide and up to 2 cm deep in the center. Nests may be initiated at any time of day but are commonly started between 1000 and 1300 or 1700 and 1800 hours. Having completed the burrow the female comes out and makes a closure by walking away from the entrance 5–10 cm, scraping sand behind her; this procedure is repeated several times, sometimes producing one or more shallow holes. Frequently females were seen to scrape a little soil for the closure, reenter, come out shortly and scrape more soil, then reenter. As many as six reentries at the time of initial closure were recorded.

Accessory burrows are of irregular occurrence in this species. In various censuses of nests, we found that from a third to half of the nests had such burrows. On 15 July 1978, for example, there were 47 active nests. Of these, 26 had no accessory burrows, 13 had one, 7 had two, and 1 had three. Most accessory burrows lay at about a right angle to the true burrow and close beside it (Figs. 2-6B, 2-11). They varied in depth from 0.2 to 6.0 cm (mean = 1.5 cm, N = 21). They were never closed, although the true burrow was normally closed except when the female entered the nest with prey. (We did note a few occasions when provisioning females failed to close the nest after leaving for a hunting trip.)

Having completed and closed the nest, the female flies over the entrance in circles or figure eights about a meter in diameter. These maneuvers begin 15–30 cm high, but she gradually rises to a height of 3–5 m, still circling, before flying off. It is noteworthy that these orientation flights take the females to the height of the male patrolling swarm.

Burrows are proclinate and quite long, up to about 75 cm, but tend to level off 25–30 cm beneath the surface and run parallel to the soil surface for much of their length. Burrow diameter is about 8 mm. The oblique burrow forms an angle of about 30° with the surface, but it soon steepens before leveling off; the extended burrow may at several points rise abruptly for 2–5 cm before dippling downward again (Fig. 3-16). It is probable that each female maintains but a single nest for her lifetime. Records for four nests show occupancies of 17 and 18 days' duration. The maximum number of cells definitely associated with one nest was 13. Because of the close proximity of nests, it was often difficult to be certain which cell was associated with a particular burrow; the 17 cells per nest reported by Evans (1970) may have resulted from confusion of cells from two nests.

Evans (1970) reported that cells are constructed on either side of the burrow in a fairly regular manner as the burrow is gradually lengthened (progressive, serial). This may sometimes be true, but in many cases the female appears to make several cells and then close off that branch of the burrow and make a second and sometimes a third branch containing several cells (Fig. 3-16B). In some nests the newest cells were clearly closer to the entrance than older cells, but that was not always true. The distance of cells from the burrow was found to vary from 1 to 15 cm (mean = 6.5 cm, $N = 11$). Cell depth varied from 8 to 20 cm (mean = 15.3, $N = 117$), distance of cells from the nest entrance from 26 to 65 cm (mean = 46.5, $N = 26$). Cells measured 15–20 mm in height and width, 18–30 mm in length. All nest excavations were made at the Deadman's Bar site.

Provisioning the Nest. Females may begin provisioning on the same day the nest is initiated. They take most of their prey at a considerable distance from the nesting area but do sometimes hunt near the nests.

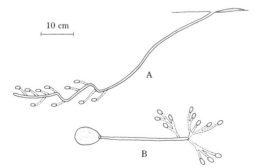

Figure 3-16. Nests of *Philanthus zebratus*, Jackson Hole, Wyo. (A) Profile of a nest with linear cell arrangement. (B) Plan view of a nest with cells in a somewhat radial pattern. Cell burrows had been filled and those drawn (dashed lines) are hypothetical.

On one occasion we tethered a bee on an *Eriogonum* in the nesting area, and within a few minutes it was seized and stung by a *zebratus* female, who attempted to carry it off. Generally, prey-laden females arrive in the nesting area at a height of 3–5 m and descend steeply to the ground, clutching the rather large prey beneath them. Sometimes they strike the ground with an audible "plop." Often they land at a considerable distance from the nest, then take a short flight to the nest, scrape it open, and enter with the prey. Frequently they drop the prey at the entrance and draw it in from the inside, often grasping the mouthparts or antennae of the prey in their mandibles.

As in other species of the genus, prey is allowed to accumulate in the burrow before being taken to a cell. Number of prey items per cell was found to vary from 3 to 9 (mean = 5.4, $N = 48$), most cells containing from 4 to 7. Prey consisted of a variety of wasps (46 species recorded; Table 3-7) and bees (32 species recorded; Table 3-8). Female bees of the genus *Osmia* constituted the single most common type of prey. Bees taken were 83% females, whereas wasps taken were only 34% females. This disparity doubtless reflects the fact that female bees spend much time on flowers, whereas female wasps tend to visit flowers more briefly and male wasps tend to spend more time there. These proportions are opposite those for the closely related species *basilaris*, however, probably reflecting a difference in hunting behavior.

Females provision with a wide range of prey shapes and sizes, sometimes bringing in prey that exceeds 75% of their own body weight. Larger females preyed upon a wider size range of prey items but generally provisioned with larger prey (head width–prey width mass correlation: $r = 0.27; N = 98; P < 0.05$).

Ferguson (1984) presented a list of prey based on material in the University of California (Berkeley) collection. All prey items were collected on 18 July 1948 at Hope Valley, Alpine County, California, by P. D. Hurd, Jr., and others. The list includes eight wasps in five genera and 28 bees in five genera. All the genera and nearly all the species are included in our list from Wyoming. As in our list, most wasps were males and most bees females.

Natural Enemies

Satellite flies were exceedingly common at the Deadman's Bar site, and prey-laden females were commonly followed by these flies, which sometimes joined them high in the air. When females land on the ground after their rapid descent, they often remain motionless for sev-

Table 3-7. Wasps taken as prey by Philanthus zebratus. (all Teton Co., Wyo.)

Species	Number taken		Family total	
ICHNEUMONIDAE			16 ♀	43 ♂
Diphyus sp.	4 ♀	10 ♂		
Dusona sp.	1 ♀			
Eutanyacra sp.	1 ♀	3 ♂		
Ichneumon spp.		12 ♂		
Pterocormus spp.	4 ♀	2 ♂		
Spilichneumon spp.	2 ♀	8 ♂		
Genus and species?	4 ♀	8 ♂		
POMPILIDAE			1 ♀	
Anoplius tenebrosus	1 ♀			
VESPIDAE: EUMENINAE			8 ♀	4 ♂
Ancistrocerus adiabatus	2 ♀	1 ♂		
Ancistrocerus catskill	2 ♀			
Eumenes crucifera nearcticus		1 ♂		
Eumenes verticalis	1 ♀			
Euodynerus annulatus		1 ♂		
Euodynerus castigatus		1 ♂		
Stenodynerus taos	1 ♀			
Symmorphus cristatus	1 ♀			
Symmorphus meridionalis	1 ♀			
VESPIDAE: MASARINAE			1 ♀	6 ♂
Pseudomasaris edwardsii	1 ♀	4 ♂		
Pseudomasaris zonalis		2 ♂		
SPHECIDAE			28 ♀	53 ♂
Ammophila azteca	3 ♀	5 ♂		
Ammophila dysmica		2 ♂		
Ammophila stangei		1 ♂		
Ammophila mediata	1 ♀			
Ammophila spp.	5 ♀	2 ♂		
Ancistromma capax		1 ♂		
Aphilanthops subfrigidus	1 ♀	6 ♂		
Astata nubecula	2 ♀			
Cerceris aequalis idahoensis	2 ♀			
Cerceris calcohorti		3 ♂		
Cerceris nigrescens		1 ♂		
Crabro latipes		1 ♂		
Crabro pleuralis	1 ♀			
Ectemnius sp.	2 ♀			
Eucerceris flavocincta		1 ♂		
Eucerceris fulvipes		2 ♂		
Oxybelus uniglumis	1 ♀			
Palmodes carbo		1 ♂		
Pemphredon sp.	1 ♀			
Philanthus pulcher	1 ♀			
Podalonia communis	2 ♀	19 ♂		
Podalonia luctuosa		3 ♂		
Podalonia occidentalis		2 ♂		
Podalonia sp.	1 ♀	1 ♂		
Tachysphex aethiops	2 ♀	2 ♂		
Tachysphex tarsatus	1 ♀			
Tachysphex sp.	2 ♀			
TOTAL 46 species	54 ♀	106 ♂		

Table 3-8. Bees taken as prey by Philanthus zebratus (all Teton Co., Wyo.)

Species	Number taken		Family total	
COLLETIDAE			9 ♀	8 ♂
Colletes consors consors		1 ♂		
Colletes kincaidii	2 ♀	4 ♂		
Colletes nigrifrons	7 ♀	3 ♂		
ANDRENIDAE			59 ♀	4 ♂
Andrena amphibola	22 ♀			
Andrena costillensis	1 ♀			
Andrena cuprotincta	1 ♀			
Andrena cyanophila	2 ♀			
Andrena medionitens	1 ♀			
Andrena milwaukeensis	5 ♀			
Andrena surda		1 ♂		
Andrena spp.	27 ♀	3 ♂		
HALICTIDAE			10 ♀	8 ♂
Agapostemon cockerelli	1 ♀			
Dufourea maura	2 ♀			
Halictus rubicundus	5 ♀	7 ♂		
Lasioglossum egregium	2 ♀			
Nomia sp.		1 ♂		
MEGACHILIDAE			118 ♀	19 ♂
Anthidium tenuiflorae		1 ♂		
Anthidium sp.	5 ♀	5 ♂		
Hoplitis fulgida	4 ♀			
Hoplitis producta interior	1 ♀			
Hoplitis sp.		4 ♂		
Megachile brevis	1 ♀	4 ♂		
Osmia calla	6 ♀			
Osmia grindeliae	4 ♀			
Osmia indeprensa	1 ♀	1 ♂		
Osmia tersula	5 ♀			
Osmia trevoris	12 ♀			
Osmia spp.	77 ♀	4 ♂		
Stelis monticola	2 ♀			
ANTHOPHORIDAE			3 ♀	3 ♂
Epeolus sp.	2 ♀	1 ♂		
Nomada sp.	1 ♀	2 ♂		
APIDAE			1 ♀	
Bombus centralis	1 ♀ (worker)			
TOTAL 32 species	200 ♀	42 ♂		

eral seconds. If a satellite fly is close by, they may fly in circuitous patterns, sometimes disappearing from sight for a minute or so. This procedure may be repeated several times if the flies persist. At times the females rise high in the air and redescend. One evasive flight lasting 5 minutes was recorded. Thus the females exhibit high approach flight and freeze-stop behavior as categorized by Alcock (1975b).

The common satellite fly was the ubiquitous *Senotainia trilineata* (Figs. 2-11, 2-12). At times as many as four flies were seen following a prey-laden female. Flies of this species were reared from maggots in cells twice. A second common miltogrammine fly was *Phrosinella pilosifrons*. These larger flies never followed females but could often be seen digging at the closure of nest entrances. They were also reared twice from maggots in cells. *Metopia argyrocephala* occurred less commonly; it was seen entering open accessory burrows on occasion but was not reared from cells. The flies are fairly successful in their attacks, as 19 of 133 cells were found to have the cell contents destroyed by maggots (14.3%).

In addition, the small conopoid flies *Zodion intermedium* and *Z. fulvifrons* were in the area and were occasionally seen perched near females. Often they perched on *Solidago* flower heads. On one occasion we saw a *Zodion* fly from a flower head onto a male *zebratus*. A struggle lasting 2–3 seconds ensued, possibly resulting in the fly depositing an egg within the body of the male.

Only one predation upon *P. zebratus* was observed. A female returning to her nest with prey landed on the web of a spider (Theridiidae). She was paralyzed by the spider and wrapped in silk.

Philanthus sanbornii Cresson

P. sanbornii occurs throughout much of the United States and southern Canada east of the Rockies. Members of the species are relatively large, comparable in size to *basilaris* and *zebratus* and slightly smaller on average than *bicinctus*. They appear to be local in distribution but often occur in large numbers in suitable areas. Most of our studies were conducted in two areas of flat, rather firm, coarse sand, both in manmade excavations, in eastern Massachusetts. An aggregation at Bedford was studied intermittently from 1966 to 1972, another at Carlisle, only about 6 km west, from 1969 to 1972. A few very brief observations were made at two other localities. A few nests were found scattered across a natural blowout on a hilltop near Manhattan, Kansas, during 1952 and 1953 (reported by Evans, 1955). Some 15–20 females were found nesting at the Archbold Biological Station, Lake Placid, Florida, in 1957 (reported by Evans and C. Lin, 1959, under the name *eurynome*, a synonym). These nests were located in a sandy field that had formerly been in agriculture but had begun growing up to short weeds. Kurczewski and Miller (1983) have also studied *sanbornii* at the Archbold Biological Station.

Throughout its range this species appears characteristically in early summer. Females were active in Florida in March and April, in Kansas in June. In Massachusetts, males and females made their appearance each year in late June (earliest date, 18 June) and had largely disappeared by late July (latest date, 25 July). Their emergence coincided closely with the flowering of *Aralia hispida* (Araliaceae), which grew in abundance in both nesting areas and was visited for nectar by *sanbornii* females and especially by the males. At Carlisle, *Ceanothus americanus* (Rhamnaceae) was also visited by numerous wasps of both sexes.

At the Massachusetts sites, we were able to make a rough estimate of the number of nests for several years. In Bedford, there were about 80 nests during the peak of the active season both in 1966 and in 1968. In 1969, the population was much higher, there being nearly 200 nests and a great many males at *Aralia* blossoms. However, in 1971 and in 1972 we were able to find only 6–10 nests. At the Carlisle site, we estimated 200 nests in 1969 and 170 in 1972. At both sites, nests in the center of the aggregation were often only 10–40 cm apart, with mounds of adjacent nests sometimes overlapping; on the periphery, nests were often more widely spaced, 1–3 m apart.

Male Behavior

After spending many hours in the nesting areas, we only twice observed males near nests. At Bedford, on 30 June 1969, we saw a male digging in the late afternoon, apparently making a burrow in which to spend the night, and on the same date a male was seen entering an abandoned burrow made by a female.

Jon Seger, of the University of Utah, has kindly sent us his notes on male behavior, which were made at an aggregation of *sanbornii* that he and J. W. Stubblefield have been studying in Littleton, Massachusetts. He reports that males emerged 3–7 days before the females and often slept in their natal burrows for several days. Sometimes groups of males, possibly brothers, continued to sleep in an old burrow for many days. At other times they made shallow sleeping burrows in the nesting area. Males left the nesting area in the morning and did not return until late afternoon. Much time was spent on the flowers of *Aralia*. Males were also seen to establish territories on a small hill about 50–70 m from the center of the nesting area. Territories were about a meter in radius and were defended from a central perch, usually a leaf or flower head. Stems on the upwind side of the territory were scent-marked as in many other species of the genus. Certain territorial sites were used day after day, often by different males. The same general area was used as the site of several territories for at least three years, and indeed one particular spot next to a *Rubus* bush is known to have been the site of a territory for all three years.

Nesting Behavior

All nests we observed were on flat or nearly flat sandy soil, sometimes close to the base of plants but more often in spaces between herbs and bushes. As she digs, the female backs from the entrance periodically, moving her abdomen up and down rapidly but not scraping sand (though she may occasionally drag out a pebble in her mandibles); when she is several centimeters from the entrance she moves back to the entrance and inside while scraping sand behind her. These movements result in the formation of several shallow grooves from the entrance and a mound of broad and irregular shape.

We found mounds at nest entrances to vary from 20 to 32 cm long by 14 to 25 cm wide. Kurczewski and Miller (1983) report mounds measuring 11 to 26 cm long (mean = 19.0 cm) and 8 to 17 cm wide (mean = 13.7 cm). Nest entrances are often quite large, as much as 25 mm wide by 15 mm high, but the burrow quickly narrows to about 8 mm in diameter. Entrances are not closed during the day, but they are closed from the inside during the night and periods of inclement weather. Mounds at nest entrances are not leveled, and there are no reports of accessory burrows in this species. Kurczewski and Miller (1983) reported much fighting between females at nest entrances. In their words:

> A female entering another's burrow was often driven backwards out of the nest, after which a period of grappling on the sand near the entrance took place. Skirmishes between conspecific females sometimes occurred when a female, backing out of her nest removing sand, was pounced upon by another female. The resident would shake off the intruder, turn to face her, and then the pair would grapple with the mandibles and forelegs. The resident would drive the intruder away and continue to remove sand from the burrow. (P. 200)

Following completion of a new nest, the female makes an orientation flight that lasts only 5–10 seconds. It begins as several small loops over the entrance and broadens to several larger loops 2–4 m in diameter and 0.5–1.0 m high. She then passes over the nest entrance and may return one or more times before flying off.

The oblique burrow penetrates the soil at a 30–60° angle, then after 12–25 cm tends to assume a greater angle. After reaching a depth of

20–30 cm the burrow frequently passes upward for 1–3 cm, then dips downward again and remains parallel with the surface or has further undulations. There are frequently lateral bends, and in some cases the burrow loops so that the cells are more or less directly beneath the entrance. The burrow is quite long, at least 70 cm and often a meter or more. Evans and C. Lin (1959) reported a burrow 1.7 m long. Kurczewski and Miller (1983) reported total lengths of 132–152 cm (mean = 144, $N = 3$) from entrance to storage area at the terminus. They found cells to be located at depths of from 77 to 85 cm (mean = 81.7, $N = 15$). These cells were very much deeper than those we studied in Massachusetts, whose depth was measured as 18.7–32.5 cm (mean = 26.2, $N = 58$) at Bedford and 16.3–30.0 cm (mean = 23.8, $N = 29$) at Carlisle.

Cells measure 12–20 mm high and wide and 25–30 mm long; they are constructed at the ends of cell burrows up to 12 mm in length (Fig. 3-17). The maximum number of cells found per nest at the Massachusetts sites was 13. The first cell is constructed close to the end of the burrow and additional cells are added progressively back toward the entrance, apparently at the rate of one a day during favorable weather. This progression of cells, which contrasts with that in *bicinctus*, was confirmed for the Florida population by Kurczewski and Miller (1983). Because of the many twists in most burrows and the fact that many cells are some distance from the burrow, it is difficult to make satisfactory excavations of mature nests; however, nests may be safely characterized as proclinate and regressive.

Provisioning the Nest. Females typically spend a few minutes clearing the entrance of their nests during the morning (0900–1030), then fly

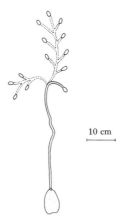

Figure 3-17. Nest of *Philanthus sanbornii*, plan view. Dashed lines show the probable route of burrows that had been filled.

10 cm

off to hunt for prey. We assume that much hunting occurs on flowers. However, Kurczewski and Miller (1983) reported capture of honey bees in front of their hives on three occasions:

> Wasps flew back and forth in front of the several-tiered hives which were located about 40 meters from the *Philanthus* aggregation. Periodically, females flew at worker honey bees as they left their hives. Prey capture involved the wasp pouncing upon the smaller bee in flight, followed by the pair falling to the ground. The wasp then stung the bee in the underside of the thorax, released it on the sand, cleaned and groomed herself, and repositioned the prey prior to transport to the nest. We did not observe any wasps taking fluids from the bees at this time. Not all honey bees were captured as they left their hives since we collected provisioning females carrying workers of *Apis mellifera* that still had much pollen on the corbiculae of the hindlegs. (P. 201)

Prey-laden females usually enter the nesting area about 0.5–1.0 m high and plunge directly into the open entrance. Occasionally, however, they descend from a height of several meters, and sometimes they drop on the soil a short distance from their nests and then fly close to the ground to the nest entrances. They are slow provisioners, often taking half an hour or more to return with prey. One female, for example, brought prey at 1045, 1111, 1200, 1324, and 1345, then closed the nest from the inside. Kurczewski and Miller (1983) reported that females spent from 11 to 77 minutes (mean = 29.2, $N = 18$) between returns to the nest. These wasps reappeared head first in their entrances from 3 to 17 minutes (mean = 6.5) after having taken their prey inside and remained in this position for 0.5–8.0 minutes (mean = 3.9) before leaving. When leaving the nest, the wasps typically walk 4–9 cm from the entrance, then take flight without further orientation.

Kurczewski and Miller (1983) found that bees in the cells had little or no pollen on their bodies, and they believe that the wasps remove the pollen before placing the bees in the cell. We have no data bearing on this matter. We did, however, sometimes find dead bees lying outside nest entrances (as we did in *bicinctus*). One female was seen to haul four bees from her nest shortly after a bout of provisioning, leaving them 4–6 cm from the entrance. We could not detect any maggots on these bees and cannot explain this behavior.

In the Massachusetts aggregations, we found the number of bees per cell to vary from 3 to 15, although most contained from 5 to 8. Those with only 3 or 4 usually contained honey bees or other large bees, while those with 10 or more were filled mainly with small halictids. Kurczewski and Miller (1983) found from 4 to 9 bees per cell; the total weight of bees in a single cell ranged from 241 to 465 mg, whereas the weight of a female was 134 mg.

Prey at the Massachusetts sites we studied was very diverse, including 32 species of bees and 5 species of wasps; honey bees comprised only 3.3% of the prey (Table 3-9). Jon Seger and William Stubblefield, of the University of Utah, have been studying a similar population in Littleton, Massachusetts, as previously mentioned (Stubblefield, 1983; J. Seger, personal communication, 1982). They have collected over 3000 prey items representing 16 families, 43 genera, and 111 species of Hymenoptera. Among the prey was a single stratiomyid fly of the genus *Odontomyia*. A detailed study of this population is in preparation by these researchers.

Kurczewski and Miller (1983) found honey bees to comprise 40% of the prey in Florida. Other bees used at the Florida site were as follows:

> Colletidae: *Colletes brimleyi*, 1 female, 24 males
> Halictidae: *Augochloropsis metallica*, 1 female
> Megachilidae: *Megachile mendica*, 1 female
> Anthophoridae: *Epeolus zonatus*, 1 female, 1 male

Honey bees have also been reported as prey of this species in Pottawatomie County, Kansas (Evans, 1955). Other bees used at this site include the halictid bees *Agapostemon radiatus* and *Lasioglossum forbesii*. At this site, also, entrances were not closed during the day, and wasps plunged directly into entrances with their prey.

Nest Duration and Final Closure. It is clear that female *sanbornii* make but a single nest during their lifetime. Of 17 nests marked on 1 and 2 July at Bedford, Massachusetts, most were still active on 12 July, but only one was still active on 23 July. No new nests were ever observed after the first week or two of the season, and marked females were never seen at a second nest. Late in the season dead females were sometimes found in inactive burrows that were still open. However, a few females were seen late in the season, apparently making final closures of their nests. These females scraped soil into the burrow from the periphery of the entrance, eventually filling it completely and leaving a series of shallow grooves radiating from the entrance.

Natural Enemies

At the Massachusetts sites, nearly every provisioning female was followed to her nest by from one to six satellite flies, *Senotainia tri-*

Table 3-9. Prey records for Philanthus sanbornii (Bedford and Carlisle, Mass.)

Species	Number taken		Family total	
TIPHIIDAE			1 ♂	
Myzinum maculatum		1 ♂		
VESPIDAE			1 ♂	
Euodynerus leucomelas		1 ♂		
SPHECIDAE				5 ♂
Aphilanthops frigidus		3 ♂		
Cerceris nigrescens		1 ♂		
Ectemnius maculosus		1 ♂		
ANDRENIDAE			21 ♀	
Andrena fragilis	1 ♀			
Andrena robertsonii	2 ♀			
Andrena spiraeana	1 ♀			
Andrena vicina	1 ♀			
Andrena wilkella	15 ♀			
Calliopsis andreniformis	1 ♀			
HALICTIDAE			101 ♀	6 ♂
Agapostemon radiatus	3 ♀	1 ♂		
Augochlora pura	1 ♀			
Augochlorella striata	3 ♀			
Augochloropsis metallica	1 ♀	1 ♂		
Dialictus coeruleus	3 ♀			
Dialictus cressonii	1 ♀			
Dialictus lineatulus	4 ♀			
Dialictus nigroviridis	6 ♀			
Dialictus spp.	7 ♀	1 ♂		
Dufourea novaeangliae		1 ♂		
Evylaeus cinctipes	1 ♀			
Evylaeus foxii	2 ♀			
Halictus confusus	22 ♀	1 ♂		
Halictus ligatus	39 ♀			
Halictus rubicundus	7 ♀	1 ♂		
Lasioglossum coriaceum	1 ♀			
MEGACHILIDAE			5 ♀	4 ♂
Chalcidoma campanulae		1 ♂		
Coelioxys modesta		1 ♂		
Heriades carinata	1 ♀			
Megachile frigida	1 ♀			
Megachile latimanus	1 ♀			
Megachile relativa		2 ♂		
Osmia sp.	2 ♀			
ANTHOPHORIDAE				2 ♂
Anthophora furcata terminalis		1 ♂		
Melissodes apicata		1 ♂		
APIDAE			5 ♀	
Apis mellifera	5 ♀ (workers)			
TOTAL 37 species (5 wasps, 32 bees)	132 ♀	19 ♂		

lineata. The wasps exhibited no escape maneuvers but plunged directly into their nests; the satellite flies often followed them in for a second or two and then emerged. Evidently maggots were successfully deposited on the prey on many occasions. Of 58 cells excavated at Bedford, 24 (41%) appeared to have had the contents destroyed by maggots. *S. trilineata* was successfully reared from a cell at Carlisle, the fly emerging 12 August from a cell excavated and containing maggots on 14 July.

Two other species of miltogrammine flies occurred less commonly at the Massachusetts sites: *Phrosinella fulvicornis* and *Metopia argyrocephala*. In addition, the mutillid wasp *Dasymutilla nigripes* was common and was often seen entering open holes, including those of *Philanthus*. The bombyliid flies *Exoprosopa fascipennis* and *Anthrax analis* were also seen ovipositing in open holes; the first of these is a known parasitoid of *Bembix* species, which were nesting nearby.

Abandoned nests at the Massachusetts sites were commonly usurped by other Hymenoptera, including ants and three species of Sphecidae that are hunters of crickets and that do not normally dig their own burrows: *Chlorion aerarium*, *Liris argentata*, and *Lyroda subita*.

In the aggregation they have been studying at Littleton, Massachusetts, J. Seger and J. W. Stubblefield (personal communication, 1982) have reared three species of parasites from cells: *Senotainia trilineata*, *Metopia argyrocephala*, and *Dasymutilla nigripes*. They have also found larvae and puparia of conopid flies in the abdomens of adult females.

Chapter 4
Species of the *gibbosus* Group

Bohart and Grissell (1975) included eight North American species in the *gibbosus* group, of which six have been studied to varying degrees and are considered here. In size, the species of this group are mostly smaller than those of the *zebratus* group but larger than those in the *politus* and *pacificus* groups. The sculpturing of the integument is relatively coarse, and the males have moderately dense, pale hairs on the abdominal venter. The two species *arizonicus* and *occidentalis* have not been studied. The remaining species are considered below in the following order: *gibbosus, crabroniformis, barbatus, multimaculatus, inversus,* and *crotoniphilus*.

Philanthus gibbosus (Fabricius)

P. gibbosus is one of the most widely distributed of North American wasps, ranging from coast to coast and from southern Canada well into Mexico. The many papers on the biology of the species vary in content but present a reasonably consistent picture of female behavior; data on male behavior are more fragmentary. Publications cover a broad geographic spectrum. Those of Evans and C. Lin (1959), Evans (1973), N. Lin (1968, 1978), Reinhard (1924, 1929), and Krombein (1952, 1956) are based on work done in the eastern states; those of the Peckhams (1898), the Raus (1918), and Barrows and Snyder (1973) are from midwestern states; Alcock (1974) studied the species in Washington, and Cazier and

Mortenson (1965) and Alcock (1978) studied it in southeastern Arizona. Our account includes additional data from Bedford, Massachusetts, from near Rensselaerville, New York, and from the vicinity of Fort Collins, Colorado. In the Rocky Mountain area, we have not found *gibbosus* nesting at elevations above 1850 m, but we have collected occasional specimens as high as 2300 m.

The typical nesting site is in bare soil of moderately firm texture, usually coarse sand, clay-sand, or sandy loam, often containing many small stones. We have found *gibbosus* nesting in paths, dirt roadways, eroded slopes, gravel pits, recently bulldozed areas, and slag piles from mines. Flat soil, slopes, and even vertical banks are utilized. In eastern Massachusetts and in central New York there is clearly only one generation a year, but near Rensselaerville, New York, there are two well-defined generations, and Reinhard (1924) reported two generations a year in Maryland. Alcock (1974) found the species nesting from mid-July to early September in Seattle, and there is undoubtedly but a single generation at that latitude, as there appears to be in Colorado. Cazier and Mortenson (1965) found a single generation in Arizona, nests being initiated after periods of summer rainfall. There are many records of wasps of both sexes visiting flowers for nectar. We have taken them on *Ceanothus*, *Aralia*, *Rhamnus*, *Leucanthemum*, *Achillea*, and *Solidago*.

Male Behavior

As first noted by the Peckhams, males commonly spend the night and periods of inclement weather in burrows occupied by females. Reinhard, N. Lin, and Evans have all confirmed this. Evans (1973) marked several males and found them to be relatively long-lived and faithful to a nest.

> In 1969, a male marked on 15 July remained active until 2 September (49 days), returning each night to the same burrow, a burrow that was inhabited in the course of the season by four different females. During 1970, several males were followed for several weeks, one of them for 47 days, returning each night to the same burrow, a burrow at times shared with other individuals of both sexes. On the other hand, males did occasionally move from one burrow to another. The maximum number of males found to emerge from any one nest in the morning was three. However, during one nest excavation, on a cool and cloudy day, a tight clump of four males was found 15 cm deep, apparently in an old, unoccupied cell. (P. 304)

The Peckhams and Reinhard have reported that males sometimes dig their own burrows in which to spend the night. Reinhard speculated that mating might occur within burrows shared with females. This seems unlikely, as recent observations suggest that the mating system of this species does not differ greatly from that of other members of the genus. Alcock (1978), working near Portal, Arizona, found males occupying perches at a cluster of yucca stalks on an earthen bank, appearing at 0800 to 0930 each morning and remaining until midday. An active nest was found within a few meters of the perch area. In his words: "Individuals perched at heights of 2–2½ m on dried branches or twigs and were separated by about 1½–3 m. The wasps regularly flew out from their perches (usually 2 to 3 times per min) for an average flight of 5.6 sec (N = 54; range 1–25 sec) before returning. Upon landing the male was likely to pump his abdomen up and down while standing on the perch or walk along the stem dragging the venter of his abdomen on the substrate for several cm" (p. 218). Alcock surmised that a pheromone was being deposited during these abdomen-dragging episodes.

He also noted males flying from their perches and pursuing interlopers. On two occasions a resident male was replaced by another individual after an interaction had taken place between them. Males held their territories for an average of about 2.2 hours per day. Only two of eight marked males appeared at a perch for more than one day. Males were removed from four perches and on the following day two of the perches were occupied by different males.

Our observations on male behavior are brief and were made at LaPorte, Colorado, on 2 September 1980. Several females nested in a steeply sloping bank. A single male had established a territory on a small willow tree, about 2.5 m tall, 10 m from the nearest known nest. He was first noted about noon and by shortly before 1400 hours had left his territory. No other males could be located and this territory was unoccupied on subsequent days.

The major perch was a horizontal leaf that extended somewhat beyond the major part of the tree's foliage; adjacent leaves were occasionally used as perches. Over a 10-minute period, 1305–1315 hours, this male changed perches 27 times, making a very short flight each time. He also made 11 longer flights, 0.3–1.0 m in length, two of them in obvious pursuit of a passing insect. Scent marking occurred along the edges of various fairly horizontal willow leaves not far from the perch; in the 10-minute period, this male scent-marked 10 times in 9 bouts. No interactions with other males were observed.

Burrow Sharing

The Peckhams (1898, p. 117) noted that "when the wasps emerge from the cocoons they find themselves in the company of their nearest relatives and in possession of a dwelling place, and they all live together for a time before starting out independently." They found a burrow early in the season containing three males and four females, which "took turns" looking out the entrance before finally leaving one by one. This behavior was confirmed by Evans (1973), who found up to three males and two females in a burrow early in the season. These associations lasted only a few days, although in several nests one male and one female occupied a burrow for as long as a month. The nests in question were in each case evidently reactivated nests of the previous generation, and it seems likely that the wasps were brothers and sisters that had recently emerged from cocoons in the nest, as assumed by the Peckhams and by Reinhard. As N. Lin (1968) has pointed out, however, there is no concrete evidence that these wasps were, in fact, siblings.

There is evidence that some nests are reused and extended by members of a second generation (see below); conceivably even a third generation might occupy the same nest, although there is no evidence for this. It appears that in some cases one of the females in the initial group of males and females takes over the nest and expands it, and in other cases the nest is simply abandoned (see Fig. 2 in Evans, 1973). Very probably this reflects the condition of the nest; that is, if it has undergone much collapse of the burrow, or if it is already a second-generation nest, it may be unsuitable for further occupation.

During the initial "sorting out" of which female would take over the nest, brief aggressive encounters sometimes occurred, and on at least two occasions the larger of two females succeeded in causing the smaller to move elsewhere (Evans, 1973). Later in the season, one rarely observes more than one female per nest. Over a 4-year period, only two such cases were observed. In one case two females were seen provisioning the same nest for one day, although the smaller of the two, for about a week, spent most of her time merely going in and out of the burrow or resting on grass blades nearby. In the second case three females were active at the same nest for several days at about midseason; for two days two of the females provisioned the nest (presumably different cells) while the third rested nearby or acted as a fortuitous guard in the entrance.

The comments above are based on observations at Bedford, Massachusetts, over a 4-year period, 1968–1971 (Evans, 1973). During the summer of 1973 these observations were continued, with very similar results. Of 20 nests active that year, 3 exhibited a turnover of females; that is, one female was replaced by another, the original owner of the nest moving elsewhere. In no case did we observe more than one female provisioning the same nest during that summer of almost continual observation.

As mentioned, one female commonly takes over a nest of the previous generation, digging out and lengthening the burrow or in some cases making a new entrance close beside the old one. Other females, probably sisters from the same nest, may start new nests close beside the old one, resulting in a distinct clump of nests (see Fig. 1 in Evans, 1973). Some entrances may be only 2–5 cm apart, with a gap of half a meter or more to the nest grouping, this general pattern being repeated year after year. The aggregation at Bedford, Massachusetts, persisted for at least 6 years, with between 20 and 30 nests each year.

Small aggregations of this nature are apparently the rule in this species (Peckham and Peckham, 1898; Reinhard, 1924). At Rensselaerville, New York, about 40 nests were scattered about in the slopes of a man-made excavation in coarse sandy loam; they tended to occur in clusters, where entrances were only 2–5 cm apart. At one site in Fort Collins, Colorado, 10 to 15 nests occurred in the sides of a steep bank of fine-grained, compacted silt; at another about 10 nests were dug into the sandy-loam slopes of a bank along an irrigation ditch. Occasional isolated nests do, however, occur.

Nesting Behavior

The digging of a new nest may require more than one day. Frequently small stones are encountered and dragged out with the mandibles. When clearing the entrance, the wasp may back out as far as 10 cm and work forward scraping soil. A typical fresh mound measures 12–15 cm long by 7–10 cm wide and about 1 cm deep in the center. One or more grooves may pass from the entrance across the mound, sometimes forming a somewhat fan-shaped tumulus. When females nest in vertical banks, however, the soil from the burrow normally falls down the slope.

There are no published reports of accessory burrows in this species, and we have noted none at sites in Colorado. However, they did occur at the study sites at Bedford, Massachusetts, and near Rensselaerville, New York. These burrows were usually made after completion of a nest but were also made in older nests, at the initial closure of the burrow in the morning (Figs. 2-6A, 4-1). On 7 August 1969, at Bedford, we found that 6 of 15 active nests had one accessory burrow each and 1 had two (the remaining 8 had none). These burrows measured from 0.5 to 5.0 cm long and lay mainly at a right angle to the true burrow. Near Rensselaerville, in 1970, we again found that about half the nests had accessory burrows, in one case as many as three. At both sites nests were in relatively coarse, hard-packed soil, and it may be that in such soil accessory burrows function as sources of soil for closure, whereas in more friable soil, accessory burrows tend to be absent because they are not needed for this purpose.

Accessory burrows are also absent in nests dug into vertical or near-vertical banks. In this instance females obtain soil for closure from around and especially just above the entrance, so that a small cave, up to 1.5 cm deep, is formed.

Nest Structure. Nests are typically proclinate, progressive, and serial. The oblique burrow may, in vertical banks, be nearly horizontal

Figure 4-1. A female *Philanthus gibbosus* opening her nest entrance (Bedford, Mass.). Note the open accessory burrow above her.

and may be straight or have small undulations or lateral bends. In all nests, at a distance of 20–30 cm from the entrance, the burrow becomes essentially horizontal. The first cells are constructed at about the point where the burrow levels off, at the ends of short cell burrows (3–5 cm long), which usually pass downward slightly from the level of the burrow. Cells are spaced 5–10 cm apart and are added as the burrow is extended farther from the entrance. Eventually the burrow may extend for more than a meter and may have as many as 17 cells (perhaps more). One two-generation nest near Rensselaerville was 150 cm long and had 15 cells with empty cocoons; beyond that there were 14 cells with eggs, larvae, or fresh cocoons (Fig. 4-2). A one-generation nest at the same locality was 160 cm long and lacked cells for the first 80 cm; beyond that there were 13 cells. Cells in any one nest tend to be at about the same depth. Indeed, cell depths throughout the range do not vary greatly, as demonstrated by Table 4-1.

In the Fort Collins area, several nests of typical structure were found in a steep sand bank northeast of the city. However, most of our work was done at the University farm, some 8 km south of this site, where the soil was loamy and often lumpy. In 1981, nests were dug into a sloping bank along an irrigation ditch, where the soil was reasonably friable, but in 1984 the only nests found were in flat, very hard-packed loam in a tree nursery. As noted in the table, cells were appreciably shallower here. In both cases the cells were very irregularly arranged, probably because of the difficulties in digging in soil of this type, and would have to be characterized as diffuse in cell pattern. One of these nests had 17 cells, with the newest cells apparently closest to the entrance.

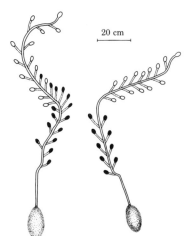

Figure 4-2. Plan view of two nests of *Philanthus gibbosus* (Rensselaerville, N.Y.). Cells of the previous generation, containing empty cocoons, are shaded; unshaded cells contained eggs, larvae, or fresh cocoons. Cell burrows, shown as dashed lines, could only be approximated. (From Evans, 1973, *Animal Behaviour*, 21: 302–308, courtesy of Bailliere Tindall, London.)

Table 4-1. Depth of cells in representative nests of Philanthus gibbosus

Locality	No. cells	Mean depth (cm)	Range (cm)
Bedford, Mass.	14	16.7	12-20
Rensselaerville, N.Y.	37	23.5	14-32
Fort Collins, Colo. (1981)	18	17.0	14-25
Fort Collins, Colo. (1984)	22	11.9	10-14

Evans and C. Lin (1959) reported cell depths varying from 17 to 30 cm at Ithaca, New York. Cazier and Mortenson (1965) described several burrows that branched, each branch having groups of cells at a different level. For example, one nest had ten cells more or less in tiers, one cell being 15.2 cm beneath the surface, two more 22.2 cm deep, two more 24.1 cm deep, three more 30.5 cm deep, the remaining two 28 and 31 cm deep. They reported the distance between cells as varying from 6 mm to several cm.

These figures all apply to nests on flat or gently sloping soil. Cells in vertical banks may be a great distance from the surface directly above. Burrow length may, as mentioned, exceed a meter but in many nests is much less than this, sometimes no more than 20–50 cm. Cell size is approximately 9 by 12 mm.

Individual nests remain active for varying lengths of time. At Bedford, some nests were active for as few as 7 days, others as many as 40 days (Evans, 1973). At Rensselaerville, several nests of the second generation, marked on 12 August, were still active on 7 September. Several nests at this site were followed over two generations; for example, one that was active on 29 June and remained so until 6 July was then inactive until 2 August, remaining active again until at least 14 August.

Orientation and Return to the Nest. Following completion of a new nest or reactivation of an old one, the female makes an orientation flight lasting several seconds. During this flight she flies from side to side in front of and facing the entrance, over an arc of increasing width (up to about a meter), then describes several loops over the entrance, only 0.5–1.5 m high, before flying off. Males orient similarly to their burrows before flying off in the morning, but their flights are usually much briefer.

Females returning to the nest with prey enter the nesting area 0.3–

1.0 m high and descend obliquely to the threshold of the nest. Characteristically, they waver from side to side during the approach flight, moving laterally 1–2 cm back and forth while descending slowly. Females followed by a *Senotainia* fly often fly beyond the entrance for a distance and may rise several meters in the air before looping back to the nest, or they may rest on vegetation some time before returning.

Reinhard (1924) also noted that females flew in a zigzag pattern when approaching the nest with prey. Alcock (1974, p. 244) reported that "as the female dropped down slowly it began at some point to move its abdomen rapidly back and forth over a narrow angle in a most pronounced way so that the wasp appeared to vibrate."

Cazier and Mortenson (1965) described the orientation flight as steplike.

> Each step consisted of from 2 to 4 hovering stops, one above the other at intervals of from 3 to 5 cm. At the end of each series of stops, the female then moves from side to side in almost a straight line for 5 to 10 cm. Each straight flight is at a slight angle away from the burrow. This is followed by another series of hovering stops. The side to side movements did not appear to increase in length as the females moved further away and as many as 15 steps were observed in a single flight. The maximum distance covered by a flight was 5.5 m vertically above the nest site. (P. 194)

These authors found that the first return flights, with or without prey, were also steplike, although later flights took the form of a "gradual downward, glide-like flight beginning from 3 to 5 m away from the entrance" (Cazier and Mortenson, 1965, p. 181). They did not observe a wavering flight similar to that reported by others.

McCorquodale (1986) described both stepwise descent and wavering movements in his studies in Alberta, Canada. In his words, the females "began a slow stepwise descent to within 1 m of their nest entrance.... Here they started a conspicuous and vigorous side-to-side waggling of their abdomen, the body held at a 45° angle to the vertical, with the head elevated, as they slowly descended directly to the nest" (p. 1623).

Provisioning the Nest. The major prey consists of ground-nesting bees that are commonly taken at their nests. The Peckhams (1898, p. 120) remarked "Our wasps had not far to go for their victims. Forty feet away... was a great *Halictus* settlement.... Into this scene of contented industry descends the ravaging *Philanthus*, taking males and females alike."

Evans and C. Lin (1959, p. 125) reported seeing a female *gibbosus* "entering the nest of a small halictid bee, apparently stinging the bees inside the nest, then carrying them directly to her own nest. This action was repeated four times in ten minutes." At Fort Collins, Colorado, we have seen *gibbosus* females attacking *Dialictus zephyrus* at two different aggregations of that bee. The wasps patrolled the bees' nesting sites constantly, from time to time seizing bees in flight and dropping to the ground to sting them. Wasps often also landed on the ground and pivoted about with their antennae extended forward, then made short flights to strike male or female bees as they flew close to the ground over the site. Brief entries into bee nests were also observed.

N. Lin (1978) also reported hunting at the nesting sites of halictid bees. Most successful attacks were made against bees leaving or returning to their nests, but 63 attacks were observed at nest entrances. In his words, "In attacking, wasps palpitate the head of the guard or the empty nest entrance with their antenna; and when the guard blocks the entrance with its abdomen, wasps have been observed attempting to grasp the bee by the abdomen with their mandibles and to pull it out. They also on occasion vigorously engage in biting away at the walls of the guarded nest entrance in attempts to gain entry. There were four distinct episodes in which wasps entered an unguarded bee nest" (p. 235).

Lin described several interactions in detail. He also reported cases in which the wasp malaxated the bee before flying off with it. According to him, under conditions of high wasp densities females establish hunting territories at bee nests. He described several intense interactions between females occupying such territories.

P. gibbosus, although by no means a specialist on one particular kind of bee, takes few bees other than Halictidae (Table 4-2). Barrows and Snyder (1973), working in Kansas, excavated 15 cells and found them to contain eight species of Halictidae, but only one specimen appeared to be *Dialictus zephyrus*, even though the wasps were nesting within an aggregation of that species. They concluded that most hunting must occur on flowers. Reinhard (1924) also reached that conclusion. It seems probable that much hunting does occur on flowers, as in most *Philanthus* species, but that certain females discover that bee nests provide a dependable source of prey and avail themselves of it.

Nest entrances are invariably closed when the female leaves. Only one to several minutes may be required by obtain prey, and the female typically remains within the nest only a few seconds to 1–2 minutes between trips. The Peckhams (1898) presented data on the time spent in these activities; one female took 13 bees in 3 hours and 9 minutes.

Table 4-2. Prey records for Philanthus gibbosus
(includes only previously unpublished records)

Species	Bedford Massachusetts	Rensselaerville New York	Larimer Co. Colorado
SPHECIDAE			
Pemphredon sp.			1 ♀
COLLETIDAE			
Hylaeus modestus	1 ♂		
Hylaeus bisinuatus	1 ♀		
Hylaeus sp.			2 ♀
ANDRENIDAE			
Calliopsis andreniformis	1 ♀		2 ♀ 1 ♂
Pterosarus andrenoides		1 ♀	
HALICTIDAE			
Agapostemon angelicus			5 ♂
Agapostemon texanus			1 ♂
Augochlora pura	10 ♀ 6 ♂	2 ♂	
Augochlorella striata	1 ♀	9 ♀ 6 ♂	1 ♀
Augochloropsis metallica	2 ♀ 1 ♂	6 ♀ 4 ♂	
Dialictus cressonii	2 ♀	1 ♀	
Dialictus delectatus	2 ♂	5 ♂	
Dialictus heterognathus		2 ♀	
Dialictus occidentalis			4 ♀ 3 ♂
Dialictus pacatus			2 ♀
Dialictus pruinosiformis			4 ♀ 17 ♂
Dialictus rohweri		5 ♀	
Dialictus ruidosensis			1 ♂
Dialictus tegulariformis			1 ♀ 1 ♂
Dialictus zephyrus		3 ♀ 1 ♂	3 ♀ 2 ♂
Dialictus spp.	1 ♂	1 ♀ 10 ♂	17 ♀ 66 ♂
Dufourea sp.			2 ♀
Evylaeus cinctipes		1 ♀	
Evylaeus pectinatus		1 ♂	
Evylaeus quebecensis	1 ♀ 1 ♂		
Evylaeus truncatus		1 ♂	
Halictus confusus	6 ♀	5 ♀ 4 ♂	4 ♀ 2 ♂
Halictus ligatus	4 ♀ 1 ♂	6 ♀ 3 ♂	
Halictus rubicundus		2 ♂	
Halictus tripartitus			1 ♀
Lasioglossum coriaceum	2 ♂	4 ♂	
Lasioglossum leucozonium		1 ♂	
Lasioglossum sisymbrii			5 ♂
Sphecodes spp.		1 ♀	1 ♀ 7 ♂
TOTALS 35 species	28 ♀ 15 ♂	41 ♀ 44 ♂	45 ♀ 111 ♂

Cazier and Mortenson (1965) reported the capture of 10 bees in 1 hour, 37 minutes, at intervals of 3–10 minutes.

There are many published prey records for *gibbosus*, from many different localities. All reveal a very strong preference for Halictidae. Only previously unpublished records are included in Table 4-2. Of the

248 records presented there, only 10 are for nonhalictid bees or for wasps. A second record for a wasp—*Crossocerus sulcus*—was presented by Evans and C. Lin (1959). Cazier and Mortenson (1965) reported males of *Colletes tucsonensis* (Colletidae) as occasional prey in Arizona, although Halictidae were more commonly used. Prey records of the Peckhams (1898), the Raus (1918), Reinhard (1924), Krombein (1952, 1956), Evans and C. Lin (1959), Alcock (1974), Cazier and Mortenson (1965), Barrows and Snyder (1973), and Packer (1985) all relate to genera of Halictidae (and many of the same species) included in Table 4-2.

The number of bees per cell varies considerably, depending on the size of the bees. We have found from 6 to 14 in Colorado; Evans and C. Lin (1959) reported from 8 to 16 in New York, Cazier and Mortenson (1965) from 8 to 23 in Arizona, Barrows and Snyder (1973) from 4 to 9 in Kansas.

Natural Enemies

Reinhard (1924) reported two species of miltogrammine flies attacking *gibbosus* in Maryland. They were *Metopia argyrocephala*, a hole searcher, and *Senotainia trilineata*, a satellite fly. He reared flies of the latter species from maggots found in cells. A species of *Senotainia* was reared from *gibbosus* cells by Barrows and Snyder (1973). Cazier and Mortenson (1965) found *S. trilineata* following female *gibbosus* in Arizona, as well as *Hilarella hilarella*, another species of satellite fly. These authors reared several *Senotainia* from infested nest cells. McCorquodale (1986) found *S. trilineata* to be a common enemy of *gibbosus* in Alberta. Of 672 approach flights he observed, 107 were followed by *Senotainia*.

S. trilineata was also common at our study sites in Massachusetts, New York, and Colorado. Evasive flights by wasps were commonly observed, as described above. At Rensselaerville, New York, a prey-laden female was seen to turn around in her nest entrance and pursue and contact a *Senotainia* that appeared about to larviposit on the prey (also noted by McCorquodale in Alberta). Near Rensselaerville, most nests had one or more cells whose contents had been destroyed by maggots (11 of 37, or about 30%). At Bedford, *Senotainia* flies were often seen trailing wasps; *Phrosinella fulvicornis* was seen digging at closed nest entrances; and *Metopia argyrocephala* was seen entering holes. The last species was reared from one cell. Of 14 cells excavated here, 5 appeared to have been attacked by miltogrammine flies (36%). At Fort

Collins, satellite flies were also prevalent; 7 of 37 cells appeared to have been attacked by these flies.

Collection of a considerable number of cocoons from a nesting aggregation of *gibbosus* at LaPorte, Colorado, in September 1978, revealed a high percentage of parasitism by the bombyliid fly *Exoprosopa fascipennis albicollaris* (determined by N. Marston). Well over half of 30 cocoons collected contained larvae of this bombyliid, and several were reared to the adult stage. This is one of a very few records of a species of *Philanthus* being attacked by bee flies.

The mutillid wasp *Dasymutilla nigripes* was abundant each year at Bedford, Massachusetts, and appeared concentrated in the *gibbosus* nesting area. On one occasion a female was seen to dig into a *gibbosus* nest and remain inside for 25 minutes. Although several *gibbosus* were reared to adulthood at this site and elsewhere, we failed to rear any mutillids. Shappirio (1948) observed *D. nigripes* entering nests of *gibbosus* in the District of Columbia, and he is doubtless correct in believing that it is a parasite of this species. In Arizona, Cazier and Mortenson (1965) reared an undescribed species of *Sphaerophthalma* (Mutillidae) from a cocoon.

At Bedford, female cuckoo wasps, *Hedychridium dimidiatum* (Chrysididae), were common in the *gibbosus* nesting area and were seen entering open burrows on several occasions.

We have observed crab spiders, *Misumena vatia*, feeding on both male and female *gibbosus* on flower heads of *Daucus carota*. Reinhard (1924) made a similar observation on *Solidago*. He also observed an asilid fly perching near a nesting aggregation in New York City and capturing prey-laden females.

At Bedford, several wasps utilized abandoned *gibbosus* burrows for nesting. These were the sphecid *Lyroda subita* and the pompilids *Ageniella conflicta* and *A. partita* (Evans, 1974).

Philanthus crabroniformis Smith

P. crabroniformis is widely distributed in the western United States. In Colorado and Wyoming we have found it chiefly in canyons and intermountain plains at elevations from about 1600 to 2200 m. Nests are located in moderately friable to rather hard-packed sandy loam, often coarse in texture and containing many stones. *P. pulcher* often nests in similar situations, and on several occasions we have found the two species nesting in much the same places. However, in the Rockies *crabroniformis* is distinctly a late summer species, appearing not earlier than late June and sometimes active into September. Thus it has scarcely begun to nest at the time when *pulcher* populations are declining. Bohart and Grissell (1975) record the species from May to September in California and suggest that there may be overlapping generations in that state. Alcock (1974) found the nesting period to extend from late June to early September in Seattle, Washington.

Our major observations were made at two localities in northwestern Wyoming. Several small aggregations were found in Jackson Hole, near the Snake River, 6.5 km west-northwest of the Moran Post Office; studies were conducted here in 1964, 1967, 1971, and 1977 (reported in part in Evans, 1970). A second aggregation occurred at Huckleberry Hot Springs, about 40 km north of the other locality; this aggregation was studied briefly in 1967 and 1977 and in more detail in 1981 (reported in part in O'Neill, 1983a). Scattered nests were also found along a trail in Hewlett Gulch, Larimer County, Colorado, in 1979 and 1981. Previously published observations include a short note by G. E. Bohart (1954) that describes females (under the name *flavifrons*, a synonym) attacking honey bees at hive entrances in California and a longer report by Alcock (1974) based on work done in Seattle, Washington.

Nests of this species tend to be aggregated. At one site in Jackson Hole, 27 nests occurred in an area measuring 8 by 8 m, with entrances separated by 10 cm to 6 m, the majority by a meter or more. At Huckleberry Hot Springs, 16 nests occurred in and beside a sandy road; 14 of them were grouped rather closely, with entrances 0.7–3.0 m apart (mean = 1.6 m). The remaining 2 nests were 8 and 10 m distant. Along the same road, 90 m east, there were 24 additional nests, also moderately clumped. At Hewlett Gulch, Colorado, several nests were scattered along a trail, entrances separated by from 0.8 to 15.0 m. Alcock (1974) worked with 40–50 nests along a hard-packed sand and gravel path through a grassy field in Seattle. At all sites, individuals of both sexes spent much time on flowers taking nectar, favored genera being *Achillea*, *Eriogonum*, *Symphoricarpos*, and especially *Solidago*.

Male Behavior

Males spend the night and periods of inclement weather in shallow burrows in the soil in and near the nesting areas of the females. In Jackson Hole, they emerged in the morning, 0930–1030, and often perched for a time on the ground or on low plants in the nesting area. During these periods they often groomed themselves, pivoted on their perches, or moved from one perch to another, without responding to conspecifics of either sex or to stones thrown by the observer. Later in the day males were often seen in abundance on flowers, but here too they failed to respond to other conspecifics of either sex.

We failed to find territorial males at the Jackson Hole site, but Alcock (1974) reported apparent territorial behavior in Seattle. One male was seen patrolling a small area and pursuing small insects; this male also walked rapidly up and down weeds, probably scent-marking. Alcock observed two copulating pairs, one of which involved a prey-laden female, although the prey was soon dropped. He described the incident as follows:

> At 14:15 on 1 July 1972, a rather small male flew behind a female carrying a prey to her nest. When she alighted in a bare spot in the middle of the path before going on to the nest entrance the male dropped down onto her back while facing in the same direction as the female. His abdomen curved downward and genital coupling occurred quickly. The female promptly took flight with the male trailing behind *in copulo* facing the opposite direction. As they passed a weed head at the edge of the path about 50 cm from the point of initial contact the male caught hold. The female tried to continue flying but, after about 5 seconds, dropped her prey and alighted. The two remained *in copulo* on the weed head for 10 minutes before separating. (P. 234)

We were able to study male behavior in some detail at the Huckleberry Hot Springs site for 8 days in July 1981. We located 15 territories 40–60 m southwest of the aggregation of 16 nests mentioned above (essentially upwind from the nests); they were dispersed along the edge of a woodland, separated by from 1 to 8 m except for one temporarily occupied territory about 20 m from the others (Fig. 4-3). There was much turnover, and five of the territories were occupied for only part of one day. Territories were only occupied when fully insolated. All were in

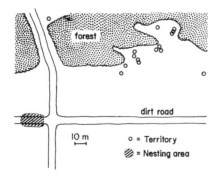

Figure 4-3. Map of study area of *Philanthus crabroniformis* at Huckleberry Hot Springs, Teton Co., Wyo.

bare spots on the ground surrounded by grasses and forbs, which were scent-marked at a high rate in the typical manner (Fig. 2-2). Freshly marked stems had a pronounced honeylike odor. In addition to occasional perching in the territory (Fig. 4-4), males spent much time hovering, especially just before scent marking. Interactions with other males usually involved a slow pursuit, one male following a few centimeters behind and slightly below the other male. Butting was frequent, grappling to the ground less frequent. Insects of other species were merely pursued a short distance.

The amount of turnover is well illustrated by the occupation of Territory 3 on 22 July. It was first occupied by an unmarked male, who was displaced by a larger male, Orange-green-orange (OGO), at 1209. This male maintained possession until 1336, when he left for no apparent reason; at 1332 he had successfully defended the territory against Blue (B), who then established a territory nearby while several times trying unsuccessfully to take Territory 3. When OGO left at 1336, B moved immediately to that territory but was displaced at 1350 by Green-white (GW), whereupon he moved back to his other territory. When GW left spontaneously at 1442, he was replaced by Y (who had previously been displaced from three other territories during the previous two hours); Y left spontaneously at 1457.

Figure 4-4. A male *Philanthus craboniformis* on a territorial perch among grasses, Teton Co., Wyo.

OGO proved to be the largest of the approximately 60 males in the area (head width, 3.8 mm). On 20 July this male successfully held Territory 4 (10 m from Territory 3) most of the day. He first occupied this territory at 1208. In a 10-minute period, from 1212 to 1222, he scent-marked 28 times in 16 bouts, marking up to three stems per flight. Most marking was on grass stems surrounding the territory, but several were on needles of a small pine nearby. During this 10 minutes, there were five interactions with intruding males, two resulting in butting and one in grappling to the ground. In the 10-minute period 1312–1322, there were 12 scent-markings in 12 bouts, nine interactions with other males, and one interaction with a *Bembix* wasp. OGO left at 1400, probably to take nectar, and returned at 1430; while he was away other males occupied this territory. At 1430 he displaced a male, and during the next 10 minutes he scent-marked four times in four bouts and interacted with other males twice. OGO left permanently at 1505, and the territory was then occupied by another male for a short time.

The importance of body size in territorial defense has been documented (O'Neill, 1983a). The head width of 63 males was found to vary from 2.4 to 3.8 mm (mean = 3.13 mm). Ten additional males were collected for analysis of dry mass, which was found to correlate strongly with head width ($r = 0.97$; $P < 0.001$). Nearly all the males in the area were marked, making it possible to record the size of winners and losers of territorial contests. Of 28 such encounters, the larger male won 27 (in the other, the two were of equal size). Of the 28 victors, 13 were territorial residents, and 15 were intruders who successfully usurped a territory.

Further confirmation of the advantage of size was obtained with removal experiments. Territorial males were removed, and the size of males that replaced them was recorded (in some cases these replacing males were also removed and the size of subsequent replacers recorded); 85% ($N = 26$) of the residents were replaced, the mean time for replacement being 8.8 minutes. Only 1 of the 22 replacing males was larger than the original resident on the territory; the remaining 21 were smaller. The mean head size of original residents was 3.40 mm, that of first replacers 2.86, a highly significant difference (t-test, $P < 0.001$). However, the size of subsequent replacers did not differ significantly from that of first replacers. The frequency of replacement by smaller males demonstrates the presence of a large group of "floaters" that move into abandoned territories whenever they are able. This practice was clearly shown in natural situations by the quick occupation of territories that had been abandoned by larger males early or late in the day or while the residents were away taking nectar.

We observed a single mating at this site, on Territory 7 at 1349 hours on 20 July. Blue-orange had been defending this territory, and when he mated with a female a smaller male took over the territory temporarily and scent-marked. The female appeared recently emerged; she had no wing wear and had but a single large egg (1.5 mm long) in her ovaries; other eggs varied in length from 0.2 to 0.7 mm.

Nesting Behavior

As mentioned above, nests are dug into coarse, moderately firm, sandy soil and are moderately aggregated. Mounds at entrances are not leveled and so are conspicuous for some time. They tend to be elongate or somewhat fan-shaped and measure 8–14 cm long by 5–14 cm wide. After some soil has accumulated, the female tends to make a groove from the entrance, and as more soil is removed she backs out this groove without scraping soil, then moves toward the entrance scraping soil behind her. Since cells are generally added daily, additional soil is added to the mound each day.

Accessory burrows are of irregular occurrence. Evans (1970) reported that at least half of the nests at the Snake River site had one or two such burrows. However, later studies at this site, on a different aggregation, showed a lower incidence of accessory burrows. In 1977, only 7 of 24 nests had one such burrow. At this site accessory burrows varied in depth from 0.3 to 2.5 cm (mean = 1.5, N = 10) (Fig. 2-6C). No accessory burrows were observed among about 40 nests at Huckleberry Hot Springs. However, at Hewlett Gulch, Colorado, all 8 nests studied had one accessory burrow; these accessory burrows varied in depth from 1 to 5 cm (mean = 3.8). Females were seen to make these burrows at the time of initial closure and to obtain soil from them for later closures. Their incidence and depth may be related to soil texture and the difficulty of finding loose soil for closure in hard substrate. The soil at Huckleberry Hot Springs was notably more sandy than at the Snake River site, and the nests at Hewlett Gulch lay along a hard-packed trail.

In a given nest, accessory burrows may vary from day to day, disappearing and then reappearing in the same or a different place. Normally they are lateral with respect to the true burrow, but in a few cases they are situated opposite the burrow, dug into the mound. Miltogrammine flies have been seen entering accessory burrows on several occasions.

The nest entrance is always closed when the female leaves the nest. At the initial closure, the female may close, reenter, and close again, sometimes as many as six times. Orientation flights consist of several loops close to the ground, about a meter in diameter, gradually increasing in height up to about 0.5 m.

Nest Structure. Nests are proclinate, progressive, and usually serial (Fig. 4-5). Burrows enter the soil at an angle of 30–45° and, after reaching a depth of 10–15 cm, tend to level off or descend at a very slight angle. Often the burrow passes abruptly upward for 2–6 cm at this point before leveling off, and there may be other undulations or lateral bends. The first cells are constructed some 18–30 cm from the entrance, but the burrow is lengthened as cells are added progressively from the entrance and may eventually reach a length of 70 cm or more. The final number of cells may be as high as 24. One female at the Snake River site was seen starting a nest on 29 July and provisioning from time to time from 1 to 26 August. When the nest was excavated on 27 August the female was still active. There were 15 cells, the newest, which contained an egg, 57 cm from the entrance. There had earlier been a separate major branch of the burrow, which had been closed off and had 9 cells, mostly with cocoons. We found one other nest that appeared to have had two successive major branches. We have no evidence that females ever make more than one nest per season.

Cells are constructed at the ends of short cell burrows (2–7 cm long). The cells lie 1–4 cm below the level of the burrow and on alternate sides (much as in *gibbosus*). Cells in any one nest tend to be at approximately the same depth. Overall cell depth at the Snake River site varied from 9 to 21 cm (mean = 15.4, $N = 77$). Excavations at Huckleberry Hot Springs yielded almost identical data (12–19 cm, mean = 15.7, $N = 13$). Cells at Hewlett Gulch were relatively close to the surface, doubtless reflecting the firmer soil at this site (8–9 cm, mean = 8.3, $N = 3$). Alcock (1974) found cell depth to vary from 12 to 16 cm at the Seattle site ($N = 24$). Cells are 10–12 mm in diameter and 14–22 mm long.

Provisioning the Nest. We observed females taking bees at flowers at two different sites. They flew from flower to flower on the downwind side, striking at various Hymenoptera and eventually succeeding in seizing and stinging a bee. Alcock (1974) also observed females going from flower to flower and darting at halictid bees, but he saw no cap-

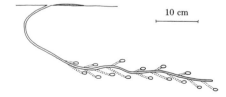

Figure 4-5. Profile of a nest of *Philanthus crabroniformis*, Teton Co., Wyo. Cell burrows had been filled and could only be approximated (shown by dashed lines). (From Evans, 1970, courtesy of Museum of Comparative Zoology, Harvard University.)

tures on flowers. Rather, he saw females taking bees at the nesting sites of ground-nesting bees, especially *Dialictus laevissimus*, as described in the following passage:

> Wasps cruised slowly a few cm above the ground frequently making right angle turns. Foraging flights (lasting 5–20 seconds in the late afternoon or on overcast days and several minutes in the middle of sunny days) were interspersed with stops during which time the wasp perched on a stone, twig, or leaf on the ground. After a brief pause, the female would fly up again. Usually, flights from a perch coincided with the appearance of other flying insects. . . . *P. crabroniformis* females pursued many species of insects including yellow jackets, bumblebees and even damselflies. Most chases ended quickly and extended over a distance of no more than 20 cm even when the insect approached was a prey species. *Dialictus* nearing their nests appeared sensitive to the presence of a female *P. crabroniformis* and often flew rapidly off in a highly erratic manner. Although the wasp, by virtue of its aerial agility, was able to orient to a bee darting back and forth, I never saw one capture a halictid that flew away from its burrow. However, if the bee did not leave but continued its final descent to the nest with a wasp slightly above and behind it, the predator would suddenly accelerate, diving at its prey with great speed. Usually the female would collide with the bee and the two would fall to the ground. There the victim was manipulated so that the wasp's jaws clamped around the "neck" of the bee; the prey was then stung in the thorax judging from one close observation of a wasp's response to a tethered bee. (P. 237)

Alcock tied a variety of objects to a strand of thread and dangled them 5 to 10 cm in front of a perched or slowly cruising wasp. The wasps could be induced to strike a wide variety of stimuli provided they were moving, were close to the ground, and were dark in color. They often attacked models of balsa that had been dipped in India ink. However, they never stung anything other than fresh bees. Alcock marked several hunting females and found that they returned frequently to the same foraging site, even as much as 3 weeks later.

Prey records for *crabroniformis* show a very strong preference for Halictidae, chiefly for species much smaller than the wasps (Table 4-3). Of 529 prey records from five localities, 94% are for Halictidae, 81% for species of *Dialictus* (which are among the smallest halictids). Thus Bohart's (1954) record of a *crabroniformis* female taking honey bees at the hive entrances is unusual, since honey bees approximate the wasps in bulk.[1] As he wrote, "It was truly astonishing to see so small a

---

1. Bohart captured this female, and it is in the collections of the U.S. Department of Agriculture Bee Biology and Systematics Laboratory at Utah State University. Its identification as *crabroniformis* has recently been confirmed by Frank D. Parker of that laboratory.

Table 4-3. Prey records for Philanthus crabroniformis

Species	Teton Co. Wyoming		Larimer Co. Colorado		Seattle Washington[1]	
SPHECIDAE						
Astata bakeri		4 ♂				
Astata nubecula	1 ♀					
Ectemnius sp.					1 ♀	
Plenoculus davisi	1 ♀					
COLLETIDAE						
Hylaeus affinis					1 ♀	
Hylaeus ellipticus	1 ♀					
Hylaeus spp.	9 ♀	2 ♂				
ANDRENIDAE						
Andrena albosellata	1 ♀					
Andrena spp.	3 ♀				1 ♀	
Perdita fallax	3 ♀					
HALICTIDAE						
Agapostemon texanus					1 ♀	
Dialictus laevissimus	47 ♀	243 ♂			23 ♀	4 ♂
Dialictus pacatus	5 ♀	2 ♂				
Dialictus perpunctatus			1 ♀			
Dialictus pruinosiformis			15 ♀	2 ♂		
Dialictus ruidosensis	1 ♀					
Dialictus spp.	48 ♀	17 ♂			4 ♀	3 ♂
Dufourea sp.	10 ♀					
Evylaeus cooleyi		1 ♂				
Evylaeus niger	1 ♀	5 ♂				
Evylaeus peraltus		2 ♂				
Evylaeus spp.	1 ♀	2 ♂	1 ♀			
Halictus confusus	1 ♀	6 ♂	1 ♀	1 ♂	3 ♀	1 ♂
Halictus farinosus					2 ♀	
Halictus rubicundus	1 ♀	6 ♂				
Halictus tripartitus	4 ♀					
Sphecodes patruelis		12 ♂				
Sphecodes sulcatulus		2 ♂				
Sphecodes spp.		15 ♂	1 ♀		1 ♀	1 ♂
MEGACHILIDAE						
Hoplitis sp.			1 ♀			
Osmia sp.	1 ♀	1 ♂				
Stelis sp.	1 ♀					
TOTALS 32 species	140 ♀	320 ♂	20 ♀	3 ♂	37 ♀	9 ♂

1. Data from Alcock (1974).

wasp ... 'beard the lions in their den'" (p. 27). His observations were described as follows:

> On June 20, 1949, at Delta, Utah, a female ... was seen in an apiary on the entrance board of a hive. She was running parallel to the entrance slot and darting at guard bees that were lined up facing her just outside the entrance. The bees backed away when the wasp closed in on them, but

four were seized in rapid succession and stung in the throat. They offered no apparent resistance and were immobilized as soon as they were stung, except for an occasional twitching of the legs. The wasp grasped the fourth victim by one antenna and had pulled her several yards across the bare ground in front of the hive when the philanthid was collected for identification. No other *Philanthus* was seen bothering the colonies in the course of brief observations made during the next few days. (P. 26)

Females bearing prey fly into the nesting area rather low, usually between 20 and 40 cm high, producing no audible sound. Usually they land on a low plant near the nest and remain motionless for several seconds before proceeding to the nest entrance. If satellite flies are present, the female may fly about and land here and there on the ground, on stems, or on the tops of plants before proceeding to the nest (Figs. 4-6, 10-1). We have seen females delayed as much as 5 minutes in entering when satellite flies are present.

Alcock (1974) made very similar observations on females' approach to the nest. He noted that the female's body is tilted slightly so that the head is higher than the abdomen as she flies slowly toward the nest. Females were often seen to pass the entrance and then to alight abruptly and "freeze" for several seconds before entering or moving to another

Figure 4-6. A female *Philanthus crabroniformis* carrying a small bee held tightly beneath her with her middle legs. This female has made a freeze stop at some distance from her nest.

perch. On one day he noted that of 38 provisioning trips recorded, 20 involved a single abrupt stop, 6 involved two, and 6 involved three or more (the remaining 6 involved no stop at all). He postulated that freezing removes the stimulus that a moving wasp evidently presents to the flies. On some occasions, flies that have been pursuing a given female will take off in pursuit of another when the initial female freezes.

Prey is brought in at irregular intervals, sometimes as little as 2 minutes being required to obtain prey, sometimes more than 30 minutes. Alcock (1974) found the average interval between trips to the nest to be about 18 minutes. The most rapid provisioning he recorded involved a wasp that brought in six bees in only 31 minutes. Since females usually provision until late afternoon, weather permitting, it seems probable that enough prey to provision a cell is normally taken per day. We found the number of prey items per cell to vary from 7 to 24 (mean = 15.3, N = 31), depending at least in part on the size of the prey; the usual number was from 12 to 18.

Despite the many resemblances between this species and *gibbosus*, we found no evidence that females shared nests with other females or with males. In 1967 we marked most of the females, and only twice did we note a female entering a nest other than her own. In one instance the intruding female had a nest only 30 cm away and apparently entered another nest by mistake, and only briefly. In another instance a female appeared to have usurped a nest that had been provisioned by a different female. Alcock noted aggressive interactions between females both at the foraging sites and at the nests. On two occasions females that approached a nest other than their own were attacked by the owner, the two grappling on the ground and the intruder soon leaving.

One nest, excavated at Huckleberry Hot Springs in 1981, was apparently an extension of a previously active *zebratus* nest. On July 19, the entrance to the nest was marked when a *zebratus* female was seen digging. This female was collected on 24 July as part of another study. On 28 July, the nest was excavated. The first three cells encountered were more than 5 cm from the main burrow, and each contained a large (*zebratus*-sized) larva that had consumed all of the prey in the cell. Two other cells were found along the next 10–12 cm of burrow and within 2 cm of the main tunnel. They were of more recent origin and contained six and eight small bees of the type and size used by *crabroniformis* as prey. These latter two cells had apparently been provisioned by the *crabroniformis* female that was sitting in the main burrow 30 cm from the nest entrance at the time the nest was excavated.

Natural Enemies

Miltogrammine flies were much in evidence in the Wyoming nesting sites during each year of the study. Satellite flies, *Senotainia trilineata*, were often seen trailing prey-laden females, eliciting the avoidance flights described above. However, they often appeared to larviposit successfully just as the female entered the nest. Two flies of this species emerged in May 1968 from cells collected in August 1967. *Phrosinella pilosifrons* was commonly seen digging at closed nest entrances, and several of these flies were also reared from cells. *Metopia argyrocephala* was observed following females and entering nests and accessory burrows, but we did not rear this species. The percentage of cells found to be infested by flies was not high. Only 20 of 94 cells contained maggots or evidence of maggots (21%). At Hewlett Gulch, Colorado, *Hilarella hilarella* was a common satellite fly.

Cuckoo wasps (*Ceratochrysis trachypleura* and *Hedychridium fletcheri*) were seen on several occasions flying from one nest entrance to another at the Wyoming sites. We have no firm evidence that this species is a parasite of *P. crabroniformis*, however.

Philanthus barbatus Smith

P. barbatus, a moderately large species, ranges from mountainous areas of the western United States well into Mexico. Along the Front Range of Colorado it is common in canyons at elevations between 1600 and 2100 m. We have found individuals of both sexes to be locally abundant on flowers of *Solidago*, *Cirsium*, and *Melilotus*. The flight period extends from late July into early September. Evans (1982) reported on a small aggregation in Hewlett Gulch, just north of Poudre Park, Larimer County, Colorado. At this site nine nests occurred along the sloping bank of a trail, and a tenth, isolated nest was dug in the bank of a stream 0.5 km away; male territories were located close beside the group of nine nests. At a second site, 23 km west of Livermore, Colorado, about 10 km from the previous site, three very widely spaced nests were studied in 1985; at this locality male territories were clustered at some distance from the nearest known nest.

Male Behavior

At Hewlett Gulch, males were found to defend territories adjacent to the site of nine nests as early as 28 July and as late as 29 August, but at no time were more than three males found to be simultaneously active. Territorial activity began at 1100–1200 hours, when the sun began to strike the site fully, and continued into midafternoon, when shade encroached on the site. Territories were 1–1.5 m in diameter. In two instances (on different dates) territories were established on the trail within 2 m of the nearest nest. Here the major perch was on the ground, and grasses 40–70 cm high beside the trail were scent-marked in a manner similar to that used by other species of the genus. Two other males were territorial about 8 m from the nests, in dense, tall grass beside the trail, where there was no bare ground. In these cases all perches were on grass blades up to 70 cm high; the perching male generally assumed an "alert" posture, facing downward but with head directed outward and antennae extended. These males made long, circling flights over the tops of grasses in and around their territories.

Males at the site 23 km west of Livermore presented a very different picture in terms of distribution. Here the three nests we were able to locate were very widely dispersed, all in sloping banks on the sides of valleys. About 15 males were found to have established territories adjacent to one another near the bottom of a narrow valley; there were also several isolated territories scattered along the lower sides of the valley. We made an extensive search for nests in this valley but found none closer than 40 m to the nearest territory.

All of these territories were in places where there was dense grass 40 to 80 cm tall. In most cases some fairly bare soil was available in the center of the territorial spaces, and males often perched on the soil or on low vegetation; in other cases there was little or no bare soil, and all perches were on grass stems well above the ground. At one site six to eight territories were adjacent to one another in a grassy slope, resulting in many interactions between the males; at another site 10 m away there were four adjacent territories, and at another, two. One of the territories was occupied on sunny days at least from 4 August through 25 August. These small groups resembled the leks of *basilaris* and as in that species were associated with widely dispersed nests. However, we also found five other territories that were, as far as we could determine, many meters from other territories and from nests.

Interactions with other males consisted of pursuit, butting in the air, and occasional grappling. Insects such as flies, butterflies, and other wasps were merely pursued from the territories. On calm days, grass stems surrounding the territory were scent-marked; on windy days, marking was chiefly on the upwind side. The number of markings per bout varied from one to four. We took 11 10-minute samples (involving

Table 4-4. Records of territorial males of Philanthus barbatus (all Larimer Co., Colo.)

Male no.	Observation period	No. scent marks Individual	No. scent marks Bouts	Interactions with conspecific males SPF[1]	Interactions with conspecific males BG[2]	Interactions with other insects[3]
1	1128-1138	17	14	3	0	1
2	1154-1204	12	8	0	0	1
3	1041-1051	33	18	1	0	0
4	1215-1225	32	28	0	0	1
5	1117-1127	37	29	0	0	5
6	1016-1026	17	8	4	2	2
	1118-1128	24	14	2	0	5
	1229-1239	19	13	1	0	2
7	1058-1108	25	15	2	1	3
	1158-1208	9	7	2	0	1
	1259-1309	8	7	0	0	1

1. Swirling or pursuit flights.
2. Butting or grappling.
3. Included 10 flies, 9 wasps, 3 butterflies.

seven individuals) and found the number of scent markings to vary from 8 to 37 (mean = 21.2) and the number of bouts to vary from 7 to 29 (mean = 14.6) (Table 4-4). As in other species, scent marking was most frequent shortly after territory occupation had begun and tended to decline with time. Perch changes were frequent, and males frequently made long, looping flights over the grass tops surrounding their territories. No matings were observed.

Nesting Behavior

Nests were without exception dug in slightly sloping sandy loam with a sparse covering of grass and forbs. As mentioned earlier, those at Hewlett Gulch were with one exception grouped in one place, nest entrances being 0.3–1.5 m apart. Nests at Livermore were, however, separated by many meters; and despite intensive searching we could find only three, in spite of the fact that the presence of a large mound at the entrance and one or more open accessory burrows makes the nests relatively conspicuous. Mounds measured 10–14 cm long by 6–16 cm wide and 1.0–1.5 cm deep in the center. Of 11 nests that were active when discovered, 3 had one accessory burrow, 7 had two, and 1 had

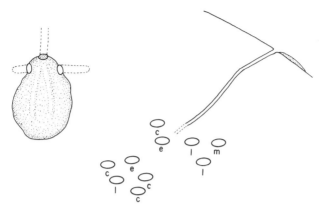

Figure 4-7. (Left) Nest entrance of *Philanthus barbatus*, showing two accessory burrows. (Right) Profile of a nest of *P. barbatus*. Cell contents: *e*, egg; *l*, larva; *m*, molded contents. (From Evans, 1982, reprinted from *The Journal of the Kansas Entomological Society*, 55(3): 571–576.)

three. These burrows were dug beside the true burrows at a 45–90° angle to it; they varied in depth from 0.2 to 5.0 cm (mean = 2.1, $N = 19$) (Fig. 4-7, left).

Burrows enter the slope at a 50–80° angle and contain a series of cells 25–40 cm from the entrance (Fig. 4-7, right). Cell burrows vary in length from 2 to 8 cm. In some nests we obtained the impression that the first cells were deep in the soil, later cells closer to the entrance; but in other nests we obtained the opposite impression. Cell depth varied from 15 to 30 cm (mean = 21.2, $N = 32$). The maximum number of cells found in one nest was 10.

Provisioning the Nest. Females were several times observed hunting at *Helianthella*, moving from flower to flower, chiefly on the downwind side. They brought prey to the nest slowly over the warmer part of the day. They entered the nesting area about a meter high, then descended slowly, obliquely to their nest entrances. Entrances were closed while the females were hunting and were opened with a few strokes of the front legs when they returned. The number of prey items per cell varied from 7 to 13 (mean = 10.3, $N = 8$). Prey consisted primarily of Halictidae of small to moderate size, but wasps of three families were also occasionally employed (Table 4-5).

No satellite flies were observed, but one of the 32 cells excavated contained several maggots, which were not reared successfully.

Table 4-5. Prey records for Philanthus barbatus (all Larimer Co., Colo.)

Species	Number taken		Family total	
TIPHIIDAE			1 ♀	
Tiphia sp.	1 ♀			
VESPIDAE			1 ♀	4 ♂
Eumenes crucifera nearcticus		4 ♂		
Stenodynerus sp.	1 ♀			
SPHECIDAE			3 ♀	
Alysson oppositus	1 ♀			
Cerceris nigrescens	1 ♀			
Passaloecus sp.	1 ♀			
ANDRENIDAE			1 ♀	
Pseudopanurgus sp.	1 ♀			
HALICTIDAE			22 ♀	58 ♂
Agapostemon texanus		7 ♂		
Augochlorella striata	3 ♀	1 ♂		
Dialictus cressonii	1 ♀			
Dialictus perdifficilis	1 ♀			
Dialictus spp.	6 ♀	6 ♂		
Evylaeus cooleyi	2 ♀			
Evylaeus spp.	2 ♀	3 ♂		
Halictus confusus	2 ♀	2 ♂		
Halictus rubicundus	2 ♀			
Lasioglossum manitouellum	1 ♀	1 ♂		
Lasioglossum sisymbrii		11 ♂		
Lasioglossum spp.	2 ♀	22 ♂		
Sphecodes sp.		5 ♂		
TOTALS 20 species (6 wasps, 14 bees)	28 ♀	62 ♂		

Philanthus multimaculatus Cameron

P. multimaculatus is a small, extensively maculated species that ranges from Alberta and British Columbia to western Texas and deep into Mexico. Bohart and Grissell (1975) characterize it as the commonest western species of the genus. Their records suggest that more than one brood may be raised per season, with a peak abundance in midsummer. In Fort Collins, Colorado, we have taken both sexes as late as 26 September, after a series of frosts had occurred. Cazier and Mortenson (1965) studied nesting several females near Portal, Arizona (under the name *P. anna*, a synonym). Alcock (1975b, 1975d) published two excellent papers based on studies also conducted near Portal. Our studies largely confirm those of Alcock. Most were conducted near Fort

Collins, Colorado, 22 July through 6 September 1978–1980. A few notes made 2 km west of Bentsen–Rio Grande State Park, Hidalgo County, Texas, on 19 May 1979 are also included.

P. multimaculatus appears to occur at low to moderate elevations and to prefer open, semiarid country. Adults have been taken on a wide variety of flowers. Dunning (1898) recorded *multimaculatus* from species of *Aster*, *Bigelovia*, and *Cleome*. In the Fort Collins area, *Eriogonum effusum* is a frequent source of nectar. Females form small, somewhat diffuse nesting aggregations in rather firm, fine-grained, compacted silt or clay-sand. The nests we studied in Colorado were in or beside a little-used dirt road closely bordered by tall grasses and weeds. They were in relatively flat soil, as was the single nest studied in Texas. However, some nests studied by Alcock were dug into the side of a vertical rodent burrow, and Cazier and Mortenson reported a preference for vertical surfaces.

Male Behavior

Alcock (1975b) found males on a low ridge partially covered with dried clumps of peppergrass (*Lepidium*), overlooking a nesting site 3 m away. Some males spent the night in burrows occupied by females, whereas others dug their own short burrows, 4 cm long; some returned repeatedly to the same burrow. By midmorning males had established perches on peppergrass plants, remaining for variable lengths of time but not past 1235 hours. Males perched in a characteristic posture, with body inclined downward but head slightly raised. Periodically they flew off to scent-mark stems in their territories in the usual V-shaped posture. These short flights were usually made into the wind. The number of flights per 5-minute period decreased significantly with the amount of time since first arrival on the perch. Alcock observed one copulation on a peppergrass stem. The female flew upwind to the stem and was dropped on quickly by the resident male. At first he mounted her facing in the same direction, then dismounted and moved to one side, the two forming a V. Duration of copulation was 5 minutes.

Alcock observed many encounters between perched males and intruders. The intruder was driven away in 32 of 37 cases. In 3 instances, however, two males were present in the same territory, one of them motionless and apparently escaping detection by the resident. Swirling flights were the most common male-male interactions; bodily contact rarely occurred.

Alcock found that ownership of perches changed frequently, either because the resident was driven off or because he left for no apparent reason. In the latter case the average occupancy was 78 minutes ($N = 7$), in the former 37 minutes ($N = 4$). As many as three different males occupied a perch in one morning. It appeared that there was a pool of nonterritorial males whose members only occasionally became territorial.

Our observations on male behavior near Fort Collins, Colorado, are similar but not identical to Alcock's. Males were present as early as 22 July and as late as 26 September, but none were found to be territorial after 6 September. In 1978, four different spaces along or close beside a dirt road where females were nesting were occupied by territorial males during the late morning hours for several days. In 1979, we found 12 territorial males along this road, one of them at a site used as a territory in 1978. Although there seemed nothing distinctive about these spaces—each was a bare area with one or more plants on the upwind side—innumerable similar spots were unoccupied. Two of the known territories were within 1.0–1.5 m of a known nest, whereas others were several meters from any known nests. Territories were 3–12 m apart.

Observations of individual males along this dirt road in 1978 and 1979 and at a sand pit in LaPorte, Colorado, in 1980 revealed that males perched either on the ground or on plant stems, either close to the ground or on top of weeds as much as 40 cm tall. However, they spent much time hovering, that is, remaining on the wing near the center of the territory either in a stationary position or moving slightly from side to side. This flight contrasts strongly with the more direct flights to stems for scent marking or toward intruders. Interactions with intruding males resembled those described by Alcock, with swirling and pursuit flights predominating. During pursuits, resident males flew slowly 2–4 cm behind and below the intruder; these flights sometimes terminated in butting or grappling. The following observations were made on one male over a period of an hour. Records on five other individuals observed over eight 10-minute periods are summarized in Table 4-6.

During the 1-hour period, 1115–1215, the male scent-marked stems 287 times, at a rate of 32 to 70 per 10-minute period. He spent much time hovering, perching on the ground or on a stick lying on the ground only occasionally, for periods of 1–10 seconds. There was no significant reduction in the number of markings per 10-minute period over the hour. Of six interactions with intruding males, all terminated in favor of the resident (no size measurements were made). All were pursuits, and one ended in grappling. An *Ammophila* and a bee fly were also pursued from the territory.

Table 4-6. Records of territorial males of Philanthus multimaculatus (all Larimer Co., Colo.)

Male no.	Observation period	No. Scent marks		Interactions with conspecific males		Interactions with other insects[3]
		Individual	Bouts	SPF[1]	BG[2]	
1	1246-1256	25	20	5	2	2
2	1120-1130	13	13	11	1	2
3	1200-1210	24	19	3	0	3
4	1211-1221	39	29	0	0	0
	1320-1330	21	15	5	2	1
	1417-1427	3	3	0	0	0
5	1155-1205	40	27	0	0	2
	1255-1305	11	11	0	0	1

1. Swirling or pursuit flights.
2. Butting or grappling.
3. Included 6 flies, 3 wasps, 1 bee, 1 beetle.

One copulation was observed, on a stem that had been marked with pheromone. The pair remained together for 3 minutes after they had first been seen. The male perched on a horizontal stem with the female suspended from his genitalia and hanging free from the stem.

Overall, these observations differ from those of Alcock in only a few details: (1) Males sometimes perched on the ground, particularly early in the day or shortly before the territories were evacuated. (2) Although interactions with other males usually consisted of swirling flights or pursuit, butting and grappling were not uncommon.

Nesting Behavior

Cazier and Mortenson (1965) presented notes on four nests near Portal, Arizona. Three of them were started in vertical surfaces, the fourth in a slope of about 5°. The researchers reported fan-shaped, unleveled mounds at nest entrances and nests left open while the females were away hunting. Alcock (1975d) found 12 nests, also near Portal, all of them under or near vegetation; several were dug into the side of a vertical rodent burrow near its entrance. Females were commonly seen digging in the morning hours (0815–1005). Alcock reported unleveled mounds and nest entrances often left open between trips for prey. Neither report mentions the presence of accessory burrows, and we have found none in our studies.

Our data on nest structure are consistent with those of Cazier and Mortenson and of Alcock. We found mounds at entrances to measure 7–8 cm long by 4–5 cm wide and about 0.5 cm deep, with no evidence of leveling. The burrow enters the soil at an angle of about 20–35° with the horizontal, then steepens considerably after a few centimeters, finally leveling off deep in the soil (Fig. 4-8A). Invariably there are sharp lateral turns in the burrow, even though the soil may be of rather uniform texture and without roots or stones. Alcock found the number of lateral turns to vary from one to six (mean = 3.8, N = 6). Our data are nearly identical (range, 1–6, mean = 2.7, N = 6). Cells are constructed from short burrows, 4–10 cm long, and are spaced 3–12 cm apart. Nests are progressive and more or less serial. Cells are 6–9 cm in diameter and 10–15 cm long. We found a maximum of six cells per nest, but because nests persist for some time (at least 18 days in one case), it is probable that more than six cells are sometimes made.

At the Colorado site, cells varied in depth from 12 to 25 cm (mean = 17.3, N = 20) and varied from 22 to 45 cm in distance from the nest entrance. Burrows in the Arizona population were somewhat longer. Alcock reported burrow lengths up to 62 cm and cells 25–35 cm deep (N = 5). Cazier and Mortenson reported burrow lengths of 35–53 cm, with cell depths of 10.8–25.4 cm. Like us, they found the newest cells deepest in the soil.

Provisioning the Nest. Females bring in small bees over the warmer parts of the day. In contrast to Arizona females, those in the Colorado population we studied always closed the entrance when leaving the nest. Alcock found that intervals between entries with prey varied from 4 to 30 minutes (mean = 12, N = 21). Comparable figures from Cazier and Mortenson are 11.5–25.5 minutes spent hunting (mean = 19.1),

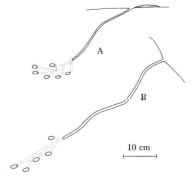

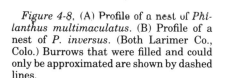

Figure 4-8. (A) Profile of a nest of *Philanthus multimaculatus*. (B) Profile of a nest of *P. inversus*. (Both Larimer Co., Colo.) Burrows that were filled and could only be approximated are shown by dashed lines.

with 1.5–2.5 minutes between each trip spent storing prey in the burrows.

Alcock found that in approaching the nest with prey, females flew rapidly toward the nest, some 0.5–1.0 m high, then dived, alighted on a bush or weed, and remained motionless for up to 30 seconds. This behavior he termed *freeze-stop*, and he regarded it as a mechanism for reducing the success of miltogrammine flies. The stops usually occurred within a meter of the nest entrance and were sometimes followed by another stop within a few centimeters of the entrance. Ultimately the females entered their nests quickly with their prey. Alcock recorded the following approaches:

1. Female shows no response (one record).
2. Female flies a few centimeters high away from the nest entrance, with one or more freeze-stops on foliage before return to the nest (six records).
3. Similar to item 2 but with noticeable changes in speed and no freeze-stops; female darts about plants before returning to the nest (six records).
4. Female flies rapidly and erratically well away from the nest, returning after a minute or more (three records).
5. Wasp flies upward and into the wind to a height of 2–6 m before turning and flying rapidly downwind, returning to the nest after about a minute (two records).

Cazier and Mortenson also reported freeze-stop behavior, and we found this behavior common in the Colorado population. Satellite flies (*Senotainia* sp.) were prevalent at the Arizona site, and this behavior appeared effective in reducing the success of the flies. Satellite flies were not common at the Colorado site, and only once was a prey-laden female seen to be followed by a fly. This female circled the area frequently, making numerous stops on low vegetation, before entering quickly without the fly in pursuit.

We found the number of bees per cell to vary from 5 fairy large ones to 23 very small ones (mean = 11.3, $N = 12$). Both in Arizona and in Colorado, the prey consisted primarily of small halictid bees (Table 4-7). The single nest we excavated in southern Texas contained a somewhat different composition of bees (not included in the table). There were 16 male *Calliopsis andreniformis*, one male *Holcopasites calliopsidis carinatus*, a female *Dialictus hunteri*, and a male *D. tegulariformis*. The *Holcopasites* is a known cleptoparasite of *Calliopsis*.

Table 4-7. Prey records for Philanthus multimaculatus

Species	Larimer Co. Colorado		Cochise Co. Arizona[1]	
SPHECIDAE				
Solierella sp.	3 ♀			
HALICTIDAE				
Dialictus clematisellus			1 ♀	
Dialictus microlepoides				1 ♂
Dialictus occidentalis	1 ♀	8 ♂		
Dialictus perparvus			9 ♀	
Dialictus pruinosiformis	6 ♀	54 ♂	8 ♀	
Dialictus pruinosus		1 ♂		
Dialictus tegulariformis		6 ♂		
Dialictus sp.			1 ♀	1 ♂
Halictus confusus	2 ♀	27 ♂		
Halictus tripartitus			11 ♀	
Sphecodes spp.	1 ♀	3 ♂	3 ♀	4 ♂
TOTALS 12 species (1 wasp, 11 bees)	13 ♀	99 ♂	33 ♀	6 ♂

1. Data from Cazier and Mortenson (1965) and Alcock (1975b).

Alcock observed females hunting on Russian thistle (*Salsola kali*). The females flew slowly around and through the plants 2–15 cm above the ground, approaching almost any moving insect. They were seen to strike at a variety of insects, including *Ammophila* much larger than themselves. Females approached within 5 cm of small bees, hovered, and then struck quickly at the prey. Alcock observed two misses and five captures. After falling to the ground, females stung their prey and then flew off carrying the prey head-forward with their middle legs.

Natural Enemies

As noted above, *Senotainia* flies were active both in Arizona and in Colorado. Alcock observed females being pursued by these flies on 9 of 17 provisioning trips, and on 3 occasions (18%) the fly appeared to make contact with the prey as the wasp reached the burrow entrance. Cazier and Mortenson watched two flies following a female within 5 or 6 cm, one of them striking the prey just outside the nest entrance. They examined the bee and found active fly larvae in its mouthparts. No data are available on the percentage of cells successfully attacked by these flies; none of the 20 cells we excavated appeared to have been parasitized.

Philanthus inversus Patton

P. inversus, a medium-sized, brilliantly patterned species, ranges through the western Great Plains to the Great Basin, sparingly into California, from southern Canada to Arizona and New Mexico. Bohart and Grissell (1975) have characterized it as an autumnal species, and we concur. In Colorado we have collected it from late July into early September. Both sexes are often taken on flowers, especially *Eriogonum effusum*, which is in bloom at that season. Females are usually found in the vicinity of steep or vertical banks of coarse sand or sandy gravel, the usual nesting sites. O'Neill and Evans (1982) have discussed prey selection in a population at Chimney Rock, Larimer County, Colorado, about 40 km southwest of Laramie, Wyoming, at an elevation of about 2350 m, and McCorquodale (1986) studied approach flights in a population at Writing-on-Stone Provincial Park in southern Alberta, Canada. Unfortunately we have no data on male behavior other than the fact that males are often seen around the nesting sites in the morning and again in the late afternoon or just before a storm. Evidently they spend the night in burrows in the banks where the females nest and spend the day well away from these sites. We excavated one sleeping burrow in a vertical bank near nests of females. It contained six males. Although we spent much time searching for male territories at Chimney Rock, we failed to find any. The females are specialized predators, and we also failed to find where they were taking prey. Quite possibly the males establish territories near sources of prey.

Nesting Behavior

In 1981 we found and marked about 30 nests in the banks of a gulley at Chimney Rock. Most nests were clumped in a limited, vertical portion of the bank, although a few were scattered in places where the slope was 30–80°. This area was adjacent to a nesting site of *bicinctus* and overlapped nests of that species; it also overlapped slightly an aggregation of *basilaris*. O'Neill and Evans (1982) reported an area of 135 m^2 occupied by *inversus*, with 30 m^2 overlap with *basilaris* and 40 m^2 overlap with *bicinctus*. Nests of *inversus* were in some cases no more than 30 cm apart, but others were separated by several meters.

Soil from these nests tended to be streaked down the bank in a linear fashion or to drop to the bottom of the bank. There was no evidence of leveling or of accessory burrows. Nest entrances were open at all times, prey-laden females plunging in directly. We excavated eight nests, all still active, and found one to five cells per nest. Considering the length of the burrows, it seems likely that each female maintains a single nest for her lifetime.

Burrows enter a vertical surface horizontally or at a slight downward angle; in slopes they make a 50–80° angle with the slope (Fig. 4-8B). All burrows angle downward more sharply after the first 15–20 cm; some angle upward sharply at certain points and then angle downward again. All burrows we excavated had one or more lateral bends ranging from gentle to sharp. Cells are constructed at the ends of cell burrows 4–6 cm long; they measure 1.5 by 3.0 cm. The oldest cells are closest to the entrance, new cells being added progressively farther from the entrance. The burrows we excavated varied in length from 40 to 64 cm, but it seems probable that in mature nests the total burrow length is considerably greater than that. Cell depth, measured from the surface of the bank, varied from 27 to 55 cm (mean = 42 cm, $N = 23$).

Provisioning the Nest. Females are specialists on male bees of the genus *Agapostemon*. Of 168 prey records from our Colorado study site, 152 were male *A. texanus* (90%). The remainder were also males: 15 *Lasioglossum* sp. and one *Colletes* sp. At this site *basilaris* used *Agapostemon* males in very small numbers, and taxonomic prey overlap with that species was very slight; there was no taxonomic or size overlap in prey with *bicinctus* and very little with *barbiger*, which also nested here. Overlap in size of prey with *basilaris* was great but of little significance because of the specialization by *inversus* on male *Agapostemon*.

David McCorquodale, of the University of Alberta, has informed us (in a personal communication, 1985) that *inversus* occurs from the third week in August to the middle of September along the Milk River in southern Alberta, where the prey consists of over 90% male *Agapostemon*.

We found the number of prey items per cell at Chimney Rock to vary from 5 to 13 (mean = 8.7, $N = 22$). Prey is, as usual, permitted to collect before being placed in a cell, but not always at the end of the burrow. For example, one nest had seven bees in a file 25 cm from the entrance but 15 cm from the end of the burrow; another had one bee 27 cm deep, three more 36 cm deep, and two at the terminus 57 cm from the entrance.

Natural Enemies

We saw no satellite flies at Chimney Rock, but 1 of 23 cells we excavated contained a single maggot (not reared successfully). McCorquodale (1986) found *Senotainia trilineata* to be a common nest parasite in southern Alberta. He followed 109 approach flights of females and found 28 of them to be followed by *Senotainia*. Approach flights by wasps were relatively slow, with from one to eight stops of 1 to 45 seconds each, generally on the ground or on plants. In six instances, during 108 approach flights, the flies made contact with the prey. Assuming eight prey items per cell, McCorquodale calculated that every third cell might have been provisioned with prey that had been contacted.

Philanthus crotoniphilus Viereck and Cockerell

We took a female *P. crotoniphilus* with prey along a moist gulley in the San Rafael Desert of Utah, 75 km southwest of Green River, at an elevation of about 1520 m, on 8 August 1981. The prey was a worker honey bee, *Apis mellifera*. Both wasp and prey had a wing length of 9.5 mm; the body length of the wasp exceeded that of the bee slightly, but because of the bee's greater bulk, its volume (measured as thorax width across the tegulae times body length) exceeded that of the wasp considerably (53.3 mm^2 compared with 34.6 mm^2). This record is of interest as one of very few records of North American species attacking honey bees.

Viereck and Cockerell (1904) described this wasp from specimens collected at Las Cruces, New Mexico, on *Croton neomexicanus*; hence the species name.

Chapter 5

Species of the *pacificus* Group

The species of the *pacificus* group are small, and all are confined to the western half of North America. In Colorado, they tend to be montane species, in contrast to species of the closely similar *politus* group, which are more characteristic of the high plains. Bohart and Grissell (1975) separated the two groups on the basis of the absence of a metapleural lamella in the *pacificus* group and its presence in the *politus* group. The behavior of only four of the nine species assigned to this group has been studied. These species are considered in the following order: *pulcher, pacificus, barbiger,* and *neomexicanus*.

Philanthus pulcher Dalla Torre

P. pulcher is widely distributed in the western United States. In Colorado and Wyoming it occurs primarily in intermountain valleys at elevations from about 1800 to 2500 m. It is characteristic of early summer; most records are from June and July. Nests are dug in moderately firm, coarse clay-loam, often containing many stones, usually in bare or sparsely vegetated places along trails or among sagebrush and grasses. Both sexes are frequent visitors to flowers near the nesting area, species of *Eriogonum* being especially favored.

Our major research site has been in Grand Teton National Park, Wyoming, more specifically in an outwash plain along the Snake River, 6.5 km west-northwest of the Moran Post Office, Teton County, at an elevation of about 2100 m. This is the same site where we studied *crabroniformis*, chiefly later in the summer. We have studied *pulcher* at this site intermittently from 1964 to 1981. Evans (1966b, 1970) has reported on nesting behavior at the site, and O'Neill (1981, 1983a) on reproductive behavior. The present report is largely a review of this information, with additional data from four other sites: (1) Hewlett Gulch, near Poudre Park, Larimer County, Colorado, at an elevation of about 2000 m; (2) 24 km west of Livermore, Larimer County, Colorado, at an elevation of 2300 m; (3) Lory State Park, near Fort Collins, Larimer County, Colorado, at an elevation of about 2000 m; and (4) Great Sand Dunes National Monument, Alamosa County, Colorado, at an elevation of about 2400 m. The Larimer County sites were in rather hard-packed loamy soil, while the Alamosa County population nested in coarse sand adjacent to the area where we studied *bicinctus* and *basilaris*.

Along the Snake River in Wyoming, females nested in dense aggregations in much the same sites over a 17-year period (Fig. 5-1). Nest entrances in the centers of these aggregations were often only 2–10 cm apart, with more scattered nests on the periphery. We could sometimes count 30–40 nests within 1 m^2; on 26 June 1979 nest entrances in one area in Jackson Hole were separated by a mean distance of 4.6 cm ($N = 39$, SD = 3.9). At one site in Hewlett Gulch, Colorado, about 50 nests were crowded along a trail in an area 2 by 3 m, with several more nests scattered farther along the trail. A second aggregation nearby had a mean internest distance of 17 cm ($N = 16$, SD = 0.13). At a third site about 1 km away, about 30 nests were distributed irregularly along the trail for some 30 m, although some nests were only 10–20 cm apart. At

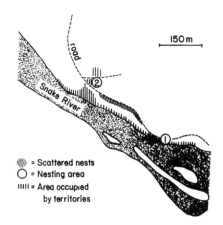

Figure 5-1. Study area for *Philanthus pulcher* along the Snake River, Jackson Hole, Wyo.

Lory State Park, 20 to 30 nests were clumped in an area 1 by 3 m along the side of a dirt road. West of Livermore, about 20 nests occupied a sloping bank over an area 3 by 6 m, entrances separated by 0.3–2.5 m. Nests appeared scattered over a wide area at Great Sand Dunes, although in one 1.0-by-2.5 m plot, seven nests were counted. Apparently, when suitably friable substrate is of uniform texture over a broader area, nests may be more scattered.

Male Behavior

Males and females appear at about the same time in mid to late June in Colorado and Wyoming, protandry appearing to be absent or slight in this species. Males spend the night in females' nests or in sleeping burrows in or near the nesting area. These sleeping burrows tend to be quite short, only 2–7 cm long, the terminus only 1–4 cm deep. A small mound no more than about 2 cm long and 1.5 cm wide is usually evident outside the entrance. In the dense aggregation in Lory State Park, we counted 25 such burrows in or near the nesting aggregation. They remained open during the day while the males were away from the nesting area. In Hewlett Gulch, there were also numerous (15–20) male sleeping burrows dispersed among the nests.

Males tend to remain in or near the nest sites until territories are established later in the morning. While in the nesting area, they perch on the ground and often pounce on one another or on females working at nest entrances. Grappling with females was commonly observed, but only once in more than 15 hours of observation (at the Wyoming site) did we see a male succeed in copulating with a female using this tactic. After coupling, the pair flew into nearby grass. However, the females generally react aggressively to these mating attempts and may try to bite the males.

The following discussion pertains to studies at the Wyoming site, with briefer notes on behavior at the Colorado sites to follow. On clear mornings, along the Snake River, males began setting up territories between 1115 and 1230, usually just before noon. By this time, most females had completed digging and were bringing in prey. Males remained territorial throughout the afternoon; the last territorial males disappeared between 1400 and 1600 hours.

Males established territories apart from nesting sites (Fig. 5-1). In 3 years of intensive study, we found only one within a nesting area, and that for only 1 day. Territories were in bare patches of soil (0.04–0.44 m^2 in area, $N = 15$) surrounded by vegetation. Territories were spaced from 1 to more than 10 m apart, chiefly along trails or in bare places among low vegetation. The mean nearest-neighbor distance between active territories on 2 days in 1978 was 4.1 m (SD = 3.6, N = 50). Most were within about 5 m of the river, where there was no appreciable concentration of flowers that might have served as major sources of nectar or prey for females. A large patch of flowering *Eriogonum umbellatum* never contained male territories. Evidently territories were chiefly in flyways used by females going to and from the nest sites. Territorial behavior was synchronized with the daily period in which females were hunting (Fig. 5-2), and the prey they took consisted of Hymenoptera, which were abundant around and beyond the territorial areas.

Males appeared to be attracted to places where the soil was relatively pale. When we artificially darkened several territorial sites with dark soil, they were abandoned. At the height of the season, when males were present in great numbers, several were seen to have established territories on pads of white, matted "cotton" from cottonwood trees (*Populus* sp.), which had accumulated among tall vegetation. We were able to create "artificial territories" by clearing small spaces and providing a surface of pale sand; such spaces were often accepted as territorial sites by males. One territory is known to have been occupied each day for 15 days, by no means always by the same male. At least two territorial perches were occupied in successive years. On the other hand, some territories were occupied only occasionally.

The males scent-marked stems surrounding the territorial perches

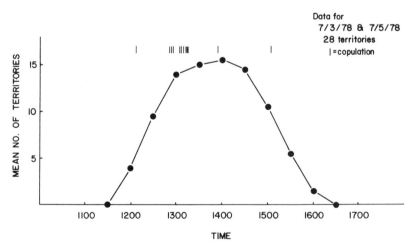

Figure 5-2. Mean number of territories of *Philanthus pulcher* on 2 days at Jackson Hole, Wyo., with times of observed copulations shown.

by passing up and down (sometimes only up) along several centimeters of the stem with head and abdomen pressed closely to it. Males perched facing upwind and made weaving flights into the wind before marking stems upwind of the perch. When the wind changed direction, males changed their orientation to face into it. In more than 7.1 hours of observation of five males, we found that a mean of 1.1 stems were scent-marked per minute. The rate of scent marking remains relatively constant across each day (O'Neill, 1981).

Responses to intruding males consisted of swirling flights, butting, and grappling. Swirling flights did not involve contact, and in no case did they result in usurpation of a territory. Twenty-six percent of the interactions in Jackson Hole consisted of swirling flights only. In 1978 and 1979, 169 responses to intruders of other species were observed. These intruders included members of six orders of insects and one spider. Only two of the responses resulted in contact with the intruder. Flying insects were rarely pursued farther than 20–30 cm (the maximum recorded was 70 cm). Males occasionally dropped on intruders on the ground.

The mean duration of occupation for 13 males that were able to successfully defend territories on warm, clear days was 73.2 minutes (SD = 60.7). The duration of occupation for six males whose territories were usurped was 17 minutes (SD = 20.9; t-test, $P < 0.01$). A given territory was frequently occupied by several males in the course of the day. The following notes on Territory 1, observed for 1 hour on 30 June 1978, serve as an illustration:

> 1234–1240: Occupied by Blue-orange, who scent-marked 15 times and undertook three swirling flights and one grapple with an intruding male.
> 1240–1246: Unmarked male usurped from Blue-orange, scent-marked 12 times, made two swirling flights and one grapple with another male, and pursued one bee and one *Ammophila* wasp.
> 1246–1334: White-white, a larger male, took over territory and in 48 minutes scent-marked 75 times, undertook 27 swirling flights and 10 grapples with conspecific males, and pursued bee flies and chrysidid wasps four times.

At 1358 White-white left the territory during a light rain. On the following day he defended a territory 5 m away, and an unmarked male defended Territory 1 for much of the day. On the third day White-white returned to Territory 1 for a short time, then moved to a territory 6 m away.

Significance of Body Size. In Jackson Hole from 1978 to 1980, 349 males were captured, measured to the nearest 0.1 mm (head width), marked with colored spots according to their size, and released. The markings permitted identification of size without recapture. In 1980, the mean size of males occupying territories was found to be 2.55 mm (SD = 0.17, N = 190). That was significantly greater than the mean for all marked males, 2.33 mm (SD = 0.22, N = 262; t-test, $P < 0.001$). The large number of marked males allowed us to determine winners and losers of prolonged aggressive interactions. In 68 recorded interactions, the larger of the two males won 61 times; in 5 cases the two were of equal size, and the smaller male won only twice. Thus, in contests between males of different size, the larger male won 97% of the time. In 57 of the 68 interactions, it was the resident that won. That appeared to reflect not resident advantage but the fact that larger males took over most territories near the beginning of the territorial period each day. Thus, once a male has established a territory, most of his subsequent fights are with smaller, nonterritorial males.

To test this hypothesis, we removed resident males from territories and recorded the size of replacement males, as well as the time between removal and replacement. The results suggested that there was a large pool of nonterritorial males. In prolonged observations in which we followed individual territories until no replacements appeared, we found the mean number of replacers per territory to be 1.4 (SD = 1.4, N = 27). The maximum number of replacers to occupy a single territory in one day was five. The mean interval between removal and replacement was 7.9 minutes (SD = 8.9, N = 38); 70% of the original residents removed were replaced. The size of first replacers was significantly less than that of original residents; first replacers' mean head width was 2.27 mm (SD = 0.11, N = 19) compared with 2.54 mm (SD = 0.13, N = 27) for original residents (t-test for significance, $P < 0.001$). However, there was no significant difference in size between first and second replacers (mean head width of second replacers was 2.28 mm, SD = 0.17). Of 36 replacers, 34 were smaller than the original resident, 1 was larger, and 1 was of equal size. Clearly, many males are excluded from territories for much of the time; presumably these floaters occupy territories after the residents have left or attempt to copulate by stealth.

Mating Behavior. Thirteen matings were observed. The duration of copulations that we saw from start to finish varied from 160 to 420 seconds and averaged 5 minutes, 7 seconds (SD = 102 seconds, $N = 7$). All but one of these matings occurred in the territorial area. In five cases, we saw the female before she entered the territory on which the mating occurred. In all five cases, she flew upwind and landed in the territory. Mating was initiated by the male straddling the female dorsally, making genital contact; then the pair quickly faced in opposite directions and usually moved to a plant on the edge of or near the territory (Fig. 5-3).

Twice a second male and once two other males attempted to disrupt a mating pair, although none was successful. On another occasion, a male entered and took possession of the territory while the previous resident was copulating. The original resident drove off the usurper after mating had been completed.

On one occasion we saw a marked, resident male and another male trying to disrupt a copulating pair on the resident's territory. Presumably the female had been attracted to the territory by the resident's chemical signals, but it was a smaller, nonresident male that succeeded in mating with her. On another occasion we observed mating by a male that was not the original resident but a replacement resulting from a removal experiment. These two cases demonstrate that floaters may, in fact, obtain occasional copulations.

Figure 5-3. A mating pair of *Philanthus pulcher* on a plant at the edge of a territory, Jackson Hole, Wyo. The female is on the left.

Data from Other Localities. The establishment of small territories that are scent-marked and defended was noted in all of the four additional localities listed above. At Great Sand Dunes, two territories were located 15–50 m away from the nesting area, a third territory actually in the middle of the rather diffuse nesting area. At Lory State Park, one territory was established only 2 m from the nesting area, and several others were widely spaced along a dirt road at a considerable distance (ca. 300 m) from a dense nesting aggregation. There is no assurance that these males were associated with this nest site, although we were unable to find other nests after an extensive search of the area.

The two aggregations in Hewlett Gulch differed greatly in the concentration of nests, as noted earlier. At one site, where about 50 nests were mostly crowded in a space 2 by 3 m, we were able to find only one territorial male, this along the trail 19 m from the nest site. Between 1239 and 1249 on 1 July 1983 this male scent-marked 21 times; in the period 1309–1319 he scent-marked 13 times (always once per bout). There were no interactions with other males and only one interaction with another insect, a fly. When the site was checked 5 days later, this territory was unoccupied, but two females were nesting 1.5 m away.

At the other, much more diffuse, nesting site, three territories were found in 1981. Two of them were within the nesting area and close to active nests; the third was along the trail 10 m from the nearest nest. One copulation was seen to be initiated in one of the territories within the nesting area. At this same site in 1982, two territorial males were observed, both within a diffuse aggregation of 22 nests scattered over 30 m of trail. One territory was 80 cm from the nearest nest, the other 190 cm from the nearest nest. Mean distance between the nests at this site was 55 cm, discounting a gap of 12 m in the trail where the soil was evidently too hard-packed for nesting.

At the site 24 km west of Livermore, Colorado, at least four males had established territories in a meadow adjacent to the bank where the females nested. The meadow contained many wildflowers, on some of which females were observed hunting. Territories were in bare places surrounded by grasses and forbs. Interactions between males were similar to those observed at the Wyoming site. Observations of one male over a 10-minute period, 1130–1140, 19 June 1985, revealed 23 scent markings in 18 bouts, with 1–3 stems marked per bout.

It appears that when nests are relatively widely spaced, as at Great Sand Dunes and the second Hewlett Gulch site, males may establish territories among the nests (as in *bicinctus* and *psyche*). At most sites, however, females dig nests very close together, and in these situations it may be impossible for males to defend territories among the nests; thus these males' territories are mainly in flyways near the nest sites.

Nesting Behavior

At the dense nesting aggregation in Jackson Hole, we noted many instances of aggression between females nesting close beside one another. These fights involved much butting and grappling between females that were digging and in a few cases between a female that was digging and one carrying prey. When we took prey from provisioning females, they attacked nearby females. Although fights sometimes lasted for over 30 minutes, rarely did a clear winner emerge, and both contestants normally returned to their nests after a time. In 1979, a fight between two females at adjacent nests ended with one female sustaining severe damage to one of her wings. On 1 July 1981, two females fought intensively for more than 10 minutes, with grappling and head butting, over possession of a single nest. The larger female eventually maintained sole possession and provisioned the nest.

Females spend the night in their nests and emerge during the morning hours, usually 0900–1000. At first they spend several minutes clearing the burrow and entrance, then make a closure, then after a brief orientation flight leave to take nectar or prey. The start of new nests, as well as the final closure of completed nests, are also commonly observed during the morning hours.

Digging of a new nest (that is, of the initial oblique burrow at the end of which the prey is stored) requires 1–3 hours. A small, elongate mound measuring about 4 by 6 cm accumulates. The digging of the initial burrow is followed by a period of leveling lasting 5–30 minutes. Following leveling, the mound is well dispersed and the soil spread over an area of about 8 by 12 cm, only a fraction of a centimeter deep. No accessory burrows have been found at any of the known nesting sites. Digging and leveling may be said to consist of the following five steps, all of them quite stereotyped:

(1) Digging the oblique, initial burrow. During this step the wasp appears frequently at the entrance, scraping soil onto the mound. At times the entrance may be plugged, and at times the female may remain inactive inside the plugged entrance, particularly if the sun becomes concealed.

(2) Grooving the mound. As the burrow approaches completion, the female backs out frequently in a straight line, making a groove through the length of the mound.

(3) Closing. The female turns to the sides and away from the opening, scraping soil into it. This procedure requires 5–15 seconds and produces a fairly thorough closure. As occurs in many species of *Philanthus*, the female commonly reenters and recloses two or more times.

(4) Leveling. The wasp backs to the far end of the mound and moves toward the entrance, with frequent, zigzagging turns to the side, while scraping soil behind her; usually she passes slightly beyond the entrance before taking flight briefly and landing in front of and facing the entrance. She then backs again to the far end of the mound and repeats the leveling process. One female made about 50 such leveling movements, each requiring 15–20 seconds.

(5) Concealing. Following a period of leveling of variable duration, the female walks off in an irregular line up to 15 cm away on the side opposite the mound, scraping soil toward the entrance. This activity is repeated several times, the movements sometimes interspersed with further leveling or with short, circling flights that undoubtedly serve as orientation flights. A somewhat longer orientation flight occurs before the wasp flies off, but these flights are quite low (2–10 cm in height) and require only a few seconds.

Nest Structure. The nests are declinate, the oblique burrow forming a 30–50° angle with the surface. This burrow varies in length from 5 to 15 cm, the terminus being at a vertical depth of from 3 to 8 cm (Table 5-1; Figs. 5-4A, 5-4B). Prey is deposited at or near the terminus until several have accumulated, usually by the afternoon of the first day. Then a cell is constructed deeper in the soil and the prey removed to a cell. The egg is then laid and the burrow leading to the cell closed off. Presumably the soil from the cell burrow is displaced into the initial burrow and then

Table 5-1. Nest dimensions (cm) of Philanthus pulcher (mean and range of variation)

Locality	No. nests	Length of oblique burrow	Depth at end of oblique burrow	Depth of cells
Wyoming: Teton Co.	6	8.6 (6-15)	6.2 (4-8)	7.9 (5-10)
Colorado: Larimer Co.	12	7.9 (5-13)	4.2 (3-7)	8.8 (6-15)
Colorado: Alamosa Co.	3	9.5 (7.5-12)	4.2 (3.8-5)	12 (10-14)

Species of the *pacificus* group   125

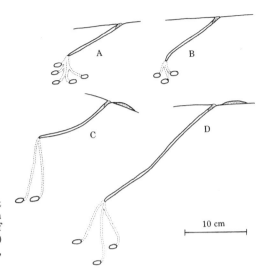

Figure 5-4. (A) A five-celled nest of *Philanthus pulcher*, Jackson Hole, Wyo. (B) A three-celled nest of *P. pulcher*, Larimer Co., Colo. (C, D) Nests of *P. barbiger*, Larimer Co., Colo.

used to close off the cell burrow, which is quite steep, often approaching 90° with the surface. The following day prey is again placed at the end of the initial burrow and later removed to a new cell. Successive cell burrows diverge from near the bottom of the initial burrow and pass off in various directions. Nests excavated one to two days after they had been started were usually found to have one to two cells, while those excavated later had more, up to seven.

The majority of nests followed had been completed and the final closure made within a week, although two were active for 10 days. Quite a number of nests became inactive after only 2 or 3 days. As noted above, females commonly begin a new nest not far from their previous nest. Since females appear to live about a month, we assume they make no more than about four nests with perhaps 10–20 cells in all.

Cells are small, about 1 cm in diameter, and are somewhat smooth-walled. Cells in any one nest tend to be at about the same depth and separated by only 2–4 cm. Cell depth was similar in the relatively firm soil in Jackson Hole and in the two sites in Larimer County, Colorado, but cells were appreciably deeper in the sandy soil at Great Sand Dunes (Table 5-1).

Provisioning the Nest. Prey consists of a wide variety of small bees and wasps (Tables 5-2, 5-3; Fig 5-5). Oddly, in Jackson Hole, of 255 prey records 38% consisted of wasps, but at the Colorado sites, of 333 prey records only 2 were wasps (0.6%). Prey-laden females enter the nesting

Table 5-2. Wasps used as prey by Philanthus pulcher

Species	Teton Co. Wyoming		Alamosa Co. Colorado	Larimer Co. Colorado
ICHNEUMONIDAE				
Genus & species?	1 ♀			
SCELIONIDAE				
Scelio sp.	1 ♀			
CHRYSIDIDAE				
Chrysura pacifica	2 ♀	1 ♂		
Elampus viridicyaneus		1 ♂		
Hedychridium fletcheri	1 ♀	4 ♂		
Holopyga ventralis	1 ♀			
Omalus aeneus	1 ♀			
VESPIDAE				
Ancistrocerus catskill		3 ♂		
Euodynerus sp.		1 ♂		
Stenodynerus papagorum	1 ♀	1 ♂		
Symmorphus canadensis	1 ♀			
SPHECIDAE				
Belomicrus coloratus				1 ♀
Belomicrus forbesii	3 ♀	9 ♂		1 ♀
Crabro florissantensis	1 ♀	1 ♂		
Dienoplius pictifrons	1 ♀	14 ♂		
Diodontus argentinae	1 ♀	2 ♂		
Diodontus gillettei		4 ♂		
Ectemnius dives		1 ♂		
Ectemnius sp.	1 ♀	2 ♂		
Lindenius columbianus	1 ♀			
Mimesa unicincta		1 ♂		
Mimumesa mixta		2 ♂		
Oxybelus uniglumis		6 ♂		
Passaloecus relativus	1 ♀	1 ♂		
Plenoculus davisi	1 ♀			
Podalonia sp.		1 ♂		
Solierella affinis		1 ♂		
Tachysphex tarsatus	4 ♀	1 ♂		
Tachysphex spp.	2 ♀	11 ♂		
Trypoxylon aldrichi		3 ♂		
TOTALS 30 species	25 ♀	71 ♂		2 ♀

area quite low, usually no more than 20 cm high, and go directly to nest entrances, which are opened with scrapes of the front legs. Wasps followed by miltogrammine flies often fly off or make irregular circling flights before entering the nest. Often they land at some distance from the nest and remain motionless for several seconds before flying to the nest and entering. Prey capture apparently often occurs on flowers, several females having been seen hovering downwind of *Eriogonum* blossoms, others at blossoms of *Phacelia*.

Table 5-3. Bees used as prey by Philanthus pulcher

Species	Teton Co. Wyoming		Alamosa Co. Colorado		Larimer Co. Colorado	
COLLETIDAE						
Colletes nigrifrons		3 ♂				
Colletes sp.						9 ♂
Hylaeus basalis	5 ♀					
Hylaeus conspicuus		2 ♂				
Hylaeus ellipticus	1 ♀	4 ♂				
Hylaeus spp.					2 ♀	8 ♂
ANDRENIDAE						
Andrena andrenoides			1 ♀			
Andrena auricoma					1 ♀	
Andrena illinoensis			2 ♀			
Andrena medionitens			2 ♀			
Andrena melanochroa	1 ♀				14 ♀	6 ♂
Andrena nasoni					1 ♀	
Andrena placida			2 ♀			
Andrena wheeleri					1 ♀	
Andrena sp.	4 ♀	9 ♂		1 ♂	2 ♀	3 ♂
Calliopsis andreniformis						1 ♂
Panurginus atriceps	7 ♀	3 ♂				
Panurginus cressoniellus		3 ♂				
Perdita wyomingensis		18 ♂				
HALICTIDAE						
Augochlorella striata					10 ♀	
Dialictus albohirtus			1 ♀			
Dialictus cressonii					13 ♀	
Dialictus laevissimus	17 ♀					
Dialictus occidentalis					1 ♀	
Dialictus pacatus					3 ♀	
Dialictus perpunctatus					2 ♀	
Dialictus pruinosiformis			1 ♀		29 ♀	1 ♂
Dialictus ruidosensis	1 ♀		2 ♀		16 ♀	
Dialictus scrophulariae			63 ♀			
Dialictus sedi					18 ♀	
Dialictus succinipennis			3 ♀			
Dialictus zephyrus					1 ♀	
Dialictus spp.	28 ♀	2 ♂			44 ♀	2 ♂
Dufourea maura	1 ♀					
Dufourea scabricornis		2 ♂				
Evylaeus pectoralis					1 ♀	
Evylaeus spp.	1 ♀				14 ♀	5 ♂
Halictus confusus	7 ♀	2 ♂			8 ♀	
Halictus tripartitus	3 ♀					
Halictus sp.			2 ♀			
Lasioglossum sisymbrii	3 ♀	1 ♂		1 ♂	14 ♀	
Lasioglossum trizonatum	1 ♀					
Sphecodes spp.	2 ♀			1 ♂	10 ♀	1 ♂
MEGACHILIDAE						
Dianthidium sp.	1 ♀	2 ♂				
Formicapis clypeata	2 ♀					
Heriades sp.						2 ♂
Hoplitis producta		1 ♂			1 ♀	1 ♂
Osmia pentstemonis	1 ♀					
Osmia spp.		13 ♂			3 ♀	1 ♂
Stelis lateralis		2 ♂				
ANTHOPHORIDAE						
Nomada spp.	1 ♀	5 ♂				
TOTALS 51 species	87 ♀	72 ♂	79 ♀	3 ♂	209 ♀	40 ♂

Figure 5-5. A female *Philanthus pulcher* entering her nest holding a pollen-laden bee beneath her with her middle legs.

Females in Jackson Hole required from 5 to 57 minutes from the time they left the nest until they returned with prey (mean = 20.5, $N = 19$). A number of females dropped the prey at the entrance, opened, and drew the prey in from the inside. Of those that dropped the prey, 70% were carrying prey larger than themselves ($N = 27$). A female at Great Sand Dunes was seen to bring in 12 bees in 100 minutes (1026–1206) and to close the nest from the inside; she resumed further provisioning at 1229 (length of provisioning flights was 6–16 minutes; mean = 9.1, $N = 11$). Females at the site near Livermore required 5–38 minutes to obtain prey (mean = 17.6, $N = 14$). Time spent within the nest between trips varied from 13 to 135 seconds (mean = 56.7, $N = 7$); these figures omit three occasions when females remained in the nest 4–11 minutes, possibly working within the burrow.

When females returned to their nests for the last time each day, they were not carrying prey, and they plugged the entrance with soil from the inside. Such females regurgitated nectar when their abdomens were squeezed slightly, indicating that their last flight was to flowers to obtain nectar. In contrast, females that were provisioning rarely regurgitated nectar when squeezed.

The number of prey items per cell was highly variable, depending in part on the size of the prey. In Jackson Hole, the number varied from 3 to 15, the usual number being 5–9 (mean = 7.5, $N = 19$). Most cells at this site contained a mixture of prey, often some bees and some wasps. At Hewlett Gulch, where most prey consisted of bees, the number of prey items per cell varied from 5 to 11 (mean = 7.5, $N = 11$); at Lory State Park the number varied from 8 to 15 (mean = 11, $N = 4$); at the locality 24 km west of Livermore, the number varied from 6 to 15 (mean = 10.8, $N = 11$).

Final Closure. Final nest closures are commonly observed during the morning hours. The female bites soil from the sides of the entrance and scrapes it into the hole. She enters periodically, head first, then backs out, turns around, and backs in scraping soil. Sometimes small lumps of earth are dragged in with the mandibles. When the burrow is fairly full, she moves away from the entrance several times, up to 5 cm away, forming a fan-shaped pattern while scraping soil toward the entrance. Final closure requires up to 45 minutes, and the result leaves the nest thoroughly concealed to a human observer.

Body Size and Nesting Success. Several lines of evidence point to an advantage in large body size among females of *pulcher*. There are significant, positive correlations of body size with both the size of eggs females carry in their ovaries and the size of prey they provision (O'Neill, 1985). Marked females at the lower end of the size range (2.0–2.3 mm in head width) were occasionally seen digging in nesting areas, but none ever provisioned nests.

Natural Enemies

In Jackson Hole, provisioning females were seen to be followed by satellite flies, *Senotainia trilineata*, on many occasions (34% of a sample of 50 prey carriage flights in 1981). We observed several apparently successful larvipositions on the prey; they occurred either as the wasp entered the nest or shortly thereafter, in which case the fly entered the burrow briefly. A larger fly, *Phrosinella pilosifrons* was also commonly seen digging at closed nest entrances, and one fly of this species was reared from a cell (emerging in May 1968 from a cell provisioned in July 1967). Despite the abundance of these flies, the percentage of infested cells at Jackson Hole was low. Of 52 cells excavated, only 2 had maggots, and 4 others appeared to have had the contents destroyed by maggots (12%). One other cell had moldy contents.

An adult female collected in Jackson Hole was found to have her abdomen filled with a fly larva of the family Conopidae, probably of the genus *Zodion*. The ovaries were pressed against the venter and were highly reduced.

On a number of occasions in Jackson Hole, cuckoo wasps, usually *Hedychridium fletcheri* but on one occasion *Ceratochrysis trachypleura*, were seen flying from one nest entrance to another but not actually entering. We have no evidence that these cuckoo wasps actually attacked *Philanthus*; there were other wasps nesting in the area, as well as several species of bees. It is interesting that cuckoo wasps were fairly frequently used as prey at this site.

At Hewlett Gulch, *Senotainia trilineata* was also abundant, and flies of this species were reared from infested cells. *Metopia argyrocephala* was also abundant and was often seen investigating open holes. A fly of this species was reared from the cell of a nest near Livermore. On one occasion a female intercepted a *Metopia* as she emerged from a nest after bringing in prey. The wasp attacked the fly by repeatedly approaching it rapidly and closely, without making contact. At Hewlett Gulch 7 of 27 cells had had the contents destroyed by maggots (26%). Near Livermore 1 of 28 cells had had the contents destroyed by maggots and two had had them destroyed by very small ants, *Solenopsis mutata*. In Jackson Hole ants carrying immobilized bees of the kind used by *pulcher* were occasionally seen in the nesting area. Females reacted aggressively to the presence of ants near their nests.

In Jackson Hole, territorial male *pulcher* were seen to be attacked by robber flies (Asilidae: *Cyrtopogon* sp.) on seven occasions. In three cases the male was killed, but in the others the male evaded the fly. These males were in the process of chasing the flies from their territories (Gwynne and O'Neill, 1980). No attacks on females were seen.

Philanthus pacificus Cresson

P. pacificus is a very small species superficially resembling *pulcher* and *psyche*. It is widely distributed west of the Rocky Mountains and occurs chiefly in areas of fine-grained sand with at most scattered vegetation. Powell and Chemsak (1959) reported on a population at dunes in Antioch, Contra Costa County, California, and Evans (1970) presented notes on females nesting along sandy roads in Jackson Hole,

Teton County, Wyoming.[1] This section is a synthesis of these reports with a few additional observations from Jackson Hole.

This species appears to have overlapping broods throughout the summer in California (Bohart and Grissell, 1975). In Wyoming it is characteristic of late July and August, appearing about the time that *pulcher* has completed nesting. We have found it nesting in small numbers on sand within the limits of the *zebratus* aggregations discussed earlier and also in alluvial sand close beside the Snake River 6.5 km west-northwest of the Moran Post Office. On one occasion two females nested only 15 cm apart and were seen to attack one another occasionally. Powell and Chemsak reported over 50 females within an area measuring about 5 by 10 m; in the place of highest concentration there were four active nests within a circle 0.5 m in diameter.

Male Behavior

Powell and Chemsak (1959) reported hundreds of males flying about roadways near the nesting site. Several of them could be seen in the morning "coursing over the burrow sites about one or two centimeters above the surface of the sand. However, during the entire day the greatest majority of males were active outside the actual nesting site" (p. 116). Several attempted matings were seen with females that were digging, but no copulations were observed.

In Jackson Hole, we several times saw males that had established territories along the sandy road and were scent-marking stems and interacting aggressively with one another (engaging in swirling flights, butting, and grappling). Unfortunately we made no detailed notes on this behavior, nor did we determine the spatial relationships of these territories to active nests.

Nesting Behavior

Powell and Chemsak (1959, p. 116) made the following observations:

> While digging, the female usually backed out, kicking the sand backward with her fore legs. Upon reaching the surface, the wasp continued to dig

---

1. The section on Sphecoidea in *Catalog of Hymenoptera in America North of Mexico* (Krombein, 1979) states that the Powell and Chemsak (1959) and Evans (1970) records are based on a misdetermination. That is not the case. In the catalog, *arizonae* Dunning is listed as a separate subspecies applicable to these records. Regardless of whether one recognizes color variants such as *arizonae* as "subspecies," *pacificus* is the species name. We do not accept *arizonae* as a valid name, nor do Bohart and Grissell (1975), who recorded both color forms from several localities.

and reentered the tunnel. After she was out of view, the digging was still evidenced by small jets of sand spurting out of the entrance, and these gradually diminished as the wasp descended. At intervals the accumulation of sand at the entrance was spread by a characteristic process. The female backed or flew three or four centimeters distant and spread the sand by kicking it back as she returned to the burrow entrance. Often this action was repeated several successive times.

Evans (1970) confirmed that the mound of soil at the entrance is leveled thoroughly, much as in *pulcher*. Up to 15 minutes may be spent in leveling. No accessory burrows have been reported for this species. An elaborate closure is made when the female leaves the nest. Powell and Chemsak (1959, p. 117) described this behavior as follows: "Beginning at the entrance, she worked her way outward for several centimeters, usually in a somewhat curving path. She kicked the loose sand back in much the same manner that she distributed it during the digging operation. This action was repeated again and again, the wasp backing or flying back to the entrance each time to start again. After the tunnel entrance was filled, the *Philanthus* female continued to rake the sand around the burrow in all directions for several centimeters."

Powell and Chemsak failed to find cells in the nests they excavated, but they did find prey stored in the burrow. Burrows were reported to be 3–4 mm in diameter and 17–22 cm long, entering the soil at an angle of about 25–30° and ending 8–12 cm below the surface. They assumed that only one cell was made per nest, but that would be unusual for a *Philanthus* and was not confirmed by our studies in Jackson Hole.

We found the oblique burrow to be 8–10 cm long and to reach a depth of 4–6 cm, where the prey was stored. From this point the wasps made a nearly vertical burrow that terminated in as many as five cells (Fig. 6-6A). Cells measured about 10 by 16 mm and varied in depth from the surface from 9 to 14 cm (mean = 11.5, $N = 11$). They were spaced 4–5 cm apart and as much as 23 cm from the entrance. Nests were clearly declinate.

Provisioning the Nest. The prey consists of a variety of small bees and wasps (Table 5-4). From 6 to 15 prey items are provided per cell (mean = 9.8, $N = 8$). Powell and Chemsak reported that provisioning was very slow, females bringing in a maximum of four prey items in 4 hours; wasps were usually gone from the nest for 40 minutes or more during foraging flights. Prey-laden females usually flew directly to the entrance, without putting down the prey or exhibiting devious flight patterns. Wasps spent from 20 seconds to 1 minute, 50 seconds within the burrow between trips for prey.

Table 5-4. Prey records for Philanthus pacificus

Species	Contra Costa Co., Calif.[1]	Teton Co. Wyoming
ICHNEUMONIDAE		
Diplazon laetatorius	1	
BRACONIDAE		
Chelonus texanus	1	
CHRYSIDIDAE		
Genus and species?		1 ♀
VESPIDAE		
Stenodynerus valliceps	1	
SPHECIDAE		
Crossocerus maculiclypeus		1 ♀
Dienoplus pictifrons		1 ♀
Diodontus spp.	11	2 ♀ 1 ♂
Lindenius columbianus		5 ♀ 1 ♂
Passaloecus relativus		1 ♀
Solierella affinis		2 ♂
Solierella blaisdelli	1	
Tachysphex sp.		2 ♂
Psenini: Genus and species?	1	
COLLETIDAE		
Hylaeus sp.		1 ♀
ANDRENIDAE		
Perdita fallax		8 ♀ 1 ♂
Perdita ciliata	2	
HALICTIDAE		
Dialictus incompletus	1	
Dialictus laevissimus		4 ♂
Dialictus ruidosensis		1 ♂
Dialictus tegulariformis	1	
Dialictus spp.	7	11 ♀ 35 ♂
Dufourea scabricornis		2 ♂
Evylaeus spp.		1 ♀ 4 ♂
Halictus confusus		2 ♂
Halictus ligatus	2	
Halictus tripartitus	2	
Sphecodes spp.	4	2 ♀ 4 ♂
TOTALS 27 species (13 wasps, 14 bees)	35	34 ♀ 59 ♂

1. Data from Powell and Chemsak (1959). These authors did not state the sex of the prey.

Natural Enemies

Powell and Chemsak collected several species of miltogrammine flies in the Antioch dunes, but they found no clear-cut evidence that any were attacking *pacificus*. We found *Senotainia trilineata* following prey-laden females at the Jackson Hole sites, but none of the cells excavated contained evidence of maggot attack.

Philanthus barbiger Mickel

P. barbiger, another small species, ranges from western Nebraska to Idaho and south to central California (east of the Sierras) and to Arizona. Bohart and Grissell (1975) characterized it as an autumnal species and reported having taken both sexes on flowers of *Chrysothamnus* (Asteraceae). Both of these statements apply well to the population in north central Colorado, where most activity occurs between mid-August and mid-September. This is the flowering period of *Chrysothamnus*, which, along with other late-season composites, appears to provide the major source of nectar for *barbiger* as well as a major source of the small Hymenoptera that serve as its prey.

We have found only one nesting aggregation of *barbiger*, at a site about 100 km northwest of Fort Collins, Larimer County, Colorado, near a geological formation known as Chimney Rock, at an elevation of about 2400 m. This area is grassland with occasional pines, junipers, and willows. The *Philanthus* occurred mainly in an area of coarse-grained, reddish soil resulting from the erosion of tall sandstone cliffs. Four species occurred here in some numbers: *barbiger*, *basilaris*, *inversus*, and *bicinctus*, all taking advantage of the late-summer abundance of flowering plants but showing little overlap in size and taxonomy of prey taken (O'Neill and Evans, 1982). Our studies were conducted between 16 August and 8 September 1977–1984, though we have collected *barbiger* as late as 15 September.

We also encountered *barbiger* males in small numbers in the San Rafael Desert of Utah (Emery County) 65 km southwest of Green River, 8–10 August 1981. The habitat was very different from the Colorado site, consisting of sandy desert at an elevation of about 1520 m. Although we collected both males and females on a variety of flowers here, we found no nests. Specimens collected here differed consistently from those from Colorado, although apparently conspecific (G. F. Ferguson, personal communication, 1983). The mesoscutum was fairly uniformly (rather than sparsely and irregularly) punctate, and both sexes were consistently larger, the males showing no overlap in size with Colorado specimens. Our data are presented in Table 5-5.

Male Behavior

This discussion pertains to the Colorado site, with data from Utah to follow. At Chimney Rock, males were found to defend territories on the

Table 5-5. Comparison of size (head width) in two populations of Philanthus barbiger

Head width (mm)	Colorado	Utah
Females	2.67-3.17 (mean = 2.89, $\underline{N}$ = 10)	2.91-3.36 (mean = 3.13, $\underline{N}$ = 10)
Males	2.06-2.60 (mean = 2.37, $\underline{N}$ = 20)	2.64-3.30 (mean = 2.98, $\underline{N}$ = 20)

ground somewhat apart from the nesting sites of the females. Territories were among short grasses, Russian thistle (*Salsola* sp.), and other plants and measured about 0.8 m in diameter. During 1977, we were able to locate only one territory, this on a slight rise about 5 m from the nearest known nest. In 1981 four territories were located, three of them about 10 m apart along a dirt road 15–25 m from a promontory overlooking a gulley, the site of several nests.

On other trips to this site territories were found in greater abundance. On 16 August 1978, we found about 30 territorial males, all at least 6 m from the nearest known nests. On 27 August 1984, 13 territories were located. On this date, as on others, territories were widely separated so that males on adjacent territories rarely interacted. Territories were spaced 2–17 m apart (mean = 6.3, SD = 5.8).

In some cases *barbiger* males established territories only a meter or two from those of the much larger *basilaris* males. In one instance three *barbiger* territories were located on a small elevation about 2.5 m in diameter, and a number of interactions between these males were observed: swirling flights, butting, and occasional grappling. Territories were occupied during the warmer part of the day, generally from about 1030 to 1300 hours, although most occupancies were quite short, often only 20–30 minutes.

Males in this population exhibited one of the highest rates of scent marking we observed in any *Philanthus*. Five males, watched over a period of 72 minutes, scent-marked at an average rate of 3.0 times per minute. Scent markings were in the usual up-down pattern, with 1–3 marks per bout. Between flights, males perched on the ground or on small stones, with none of the prolonged hovering reported for some *Philanthus* males (e.g., *multimaculatus* and a Utah population of *barbiger*, as described below). Position of the perch shifted with wind direction, so males almost invariably perched downwind of the scent-marked plants. Weaving flights prior to marking were much as described for *psyche* (Fig. 6-4).

Aggressive interactions on territories were not as common as observed in some other species (e.g., *pulcher*, *psyche*), but escalated interactions involving butting and grappling were occasionally observed. One usurpation of a territory by a larger intruder was seen. Most interactions with other insects simply involved pursuits of conspecific males as well as other insects (e.g., bees and chrysidid wasps).

Evidently there were a number of floaters in the area. We conducted several removal experiments; for example, on 21 August 1981, a male with a head width of 2.6 mm was removed from his territory at 1130 and a smaller male (head width, 2.3 mm) replaced him in 1 minute. The second male was removed at 1136 and was replaced by a male of similar size in 9 minutes. This male was in turn removed at 1146 and was replaced in 5 minutes by a male whose head width was 2.2 mm. This male was removed, but he was not replaced, as the sky had become overcast. Overall, 71% of all males removed from territories during sunny periods were replaced by other males ($N = 14$).

In the San Rafael Desert of Utah, a much hotter environment than Chimney Rock, several males were seen to have established territories on the tops of rather tall *Erigonum* plants. At a nearby site, in a depression between sand dunes, five males were found to have established territories near midday, all downwind of dead weed stalks up to a meter tall. Territories were about 1 m in diameter; four were adjacent to one another, the fifth 15 m away. Scent markings were brief, with upward but no following downward movements; all took place on the tall weeds, behind which the males spent much time hovering, occasionally perching either on the weeds or on low green plants in the territory (never on the ground).

Observations of one male over a 10-minute period revealed 24 scent markings in 16 bouts. There were two interactions with males on adjacent territories, both swirling flights with no bodily contact. During the 10-minute period, this male spent a total of about 3 minutes hovering (in 22 bouts), about 6 minutes perching, and the remaining minute scent marking and defending the territory.

Although the high frequency of scent marking resembled that in the Colorado population, these wasps spent their time well above the ground, with much hovering. This behavior may have been related to the high temperature of the sand surface in this desert area. The statement that scent marking here involves an upward and no downward movement is based on observations of a single male and requires confirmation. Indeed, further studies of these two quite different populations should prove most interesting.

Nesting Behavior

All observations on nesting behavior were made at the Colorado site. Although male territories were widely dispersed at the site, females appeared to restrict much of their nesting to a plot measuring about 3 by 8 m in flat or slightly sloping soil near the edge of a gulley. Some 10 nests were crowded on a small promontory where the soil was bare and somewhat more friable than elsewhere. Some of these nests were no more than 25 cm apart. Other nests were scattered widely along a nearby little-used dirt road.

When digging a nest, a female emerges only a short distance from the entrance, and the soil tends to accumulate in a circular mound, which remains intact through the nesting cycle. We observed no leveling movements whatever. Most females, however, dig a short accessory burrow at a 20–90° angle to the true burrow after the latter has been completed. Of 12 nest entrances measured, all had roughly circular mounds measuring 4–9 cm in diameter. Eight had one accessory burrow each; the accessory burrows measured from 0.2 to 6.0 cm long (mean = 2.7 cm). One nest had two accessory burrows, 5 and 7 cm long; the remaining nests had none at the time they were discovered. Accessory burrows were typically left open, whereas true burrows were invariably closed except when the female was actually entering or leaving. On some occasions females were seen to enter accessory burrows and remain inside for several seconds.

Nests are declinate and resemble those of *pulcher* and *pacificus* (Figs. 5-4C, 5-4D). The oblique burrow forms a 40–60° angle with the surface. During provisioning, prey items are stored at the bottom of this initial burrow, which measures 15–31 cm long, with the terminus 9–18 cm below the surface ($N = 5$). Most burrows were found to be straight, but some had one or more lateral bends, and one made a loop so that the cell was nearly beneath the entrance. Cells are constructed well below the terminus at depths of 19–26 cm (mean = 23.0, $N = 10$). Cells also tend to be well beyond the terminus, as much as 40 cm measured in a straight line from the entrance. The maximum number of cells found in any one nest was three; they were 6–10 cm apart and 20–24 cm deep. Cells were small and contained from 7 to 12 prey items ($N = 5$).

Nest duration is apparently rather short, perhaps 3 to 5 days. Nests marked on 23 August were all inactive by 8 September, but most were still visible in the form of partially eroded mounds and broad accessory burrows. We believe that the accessory burrows serve chiefly as quarries for fill in this coarse soil, and thus they may be enlarged at the time of final closure.

Provisioning the Nest. Females appear to do much hunting on the flowers of *Chrysothamnus, Gutierrezia*, and other composites. A diversity of small Hymenoptera are accepted as prey (Table 5-6). Females approach their nests with prey quite low, only 5–20 cm above the ground. In some cases the prey is virtually as large as the predator (e.g., female *Tachysphex terminatus*). O'Neill and Evans (1982) reported the mean prey size to be 10.1 mm^2 (measured as the product of thorax width across the tegulae and total body length). This was very much less than that of the three other species nesting nearby, *inversus, basilaris*, and *bicinctus*, in which mean prey size ranged from 24.2 to 52.9 mm^2. There was, however, some overlap with *inversus* and considerable overlap with *basilaris*, with respect to both prey size and prey type. Six genera and two species appear on the prey lists of both *barbiger* and *basilaris*. However, *basilaris* took significantly larger prey on average (O'Neill and Evans, 1982).

Natural Enemies

One prey-laden female was seen to be followed by a satellite fly, *Senotainia trilineata*. This female flew about in irregular circles, entering the nest after a minute or two without the fly in pursuit. One cell excavated had a single maggot that had devoured most of the cell contents. This maggot pupated soon after we had discovered it and produced an adult fly the following spring, on 18 March. The fly was identified as *Phrosinella pilosifrons*.

Philanthus neomexicanus Strandtmann

Knowledge of this species is based on a single nest excavation made in dunes at Point Reyes, California, by Gambino (1985). The burrow was 12 cm long and reached a vertical depth of 8 cm; at this depth, several prey items were found. These were *Lasioglossum pavonotum* (Halictidae) (one female, two males) and *Sphaerophoria cylindrica* (Syrphidae) (one male). The latter represents one of only two records of a North

Species of the *pacificus* group

Table 5-6. Prey records for Philanthus barbiger (all Larimer Co., Colo.)

Species	Number taken		Family total	
BRACONIDAE			1 ♀	1 ♂
Chelonus sp.	1 ♀	1 ♂		
ICHNEUMONIDAE			6 ♀	7 ♂
Anomalon sp.	4 ♀	5 ♂		
Glypta sp.	1 ♀			
Melanichneumon sp.	1 ♀			
Parania geniculata		1 ♂		
Genus and species?		1 ♂		
PERILAMPIDAE			1 ♀	
Perilampus sp.	1 ♀			
TIPHIIDAE			7 ♀	
Tiphia sp.	7 ♀			
VESPIDAE			1 ♀	4 ♂
Ancistrocerus sp.		3 ♂		
Stenodynerus sp.	1 ♀	1 ♂		
SPHECIDAE			7 ♀	
Crossocerus sp.	1 ♀			
Dryudella rhimpa	1 ♀			
Mimesa sp.	3 ♀			
Oxybelus emarginatus	1 ♀			
Tachysphex terminatus	1 ♀			
COLLETIDAE			1 ♀	4 ♂
Colletes sp.		1 ♂		
Hylaeus coloradensis	1 ♀	3 ♂		
ANDRENIDAE			4 ♀	4 ♂
Perdita sp.	4 ♀	4 ♂		
HALICTIDAE			15 ♀	23 ♂
Dialictus albohirtus	4 ♀			
Dialictus lazulis	3 ♀	1 ♂		
Dialictus ruidosensis	1 ♀	2 ♂		
Dialictus sp.		3 ♂		
Evylaeus ovaliceps	2 ♀			
Evylaeus sp.	1 ♀	1 ♂		
Lasioglossum egregium		1 ♂		
Lasioglossum sisymbrii		2 ♂		
Sphecodes spp.	4 ♀	13 ♂		
ANTHOPHORIDAE			1 ♀	
Nomada sp.	1 ♀			
TOTAL 28 species (15 wasps, 13 bees)	44 ♀	43 ♂		

American *Philanthus* using a fly as prey. This fly has a banded pattern not unlike that of the bees found in the burrow. Gambino failed to find cells in this nest and pointed out that the female may have rejected the fly before placing it in a cell; it is also possible that the larva might have refused it as food.

Chapter 6
Species of the *politus* Group

The *politus* group includes the most diminutive of all *Philanthus*; the smallest species, *parkeri*, measures only 5–8 mm in length. We prefer to include only five species in this group. Ferguson (1983) also included *albopilosus* and *ventilabris*. Although it is true that these two species have a metapleural lamella, as do *politus* and closely related species, they have other structural peculiarities, and we prefer to consider them separately, in Chapter 7. Males of the *politus* group have moderately dense, pale hairs on the abdominal venter and all (so far as studied) scent-mark their territories. The five species we include in this group are considered in the following order: *psyche, tarsatus, serrulatae, politus*, and *parkeri*.

Philanthus psyche Dunning

P. psyche is locally abundant in broad areas of fine-grained sand in the western Great Plains and Great Basin, from Alberta and the Dakotas to Utah and western Texas. In the dunes where it occurs, the common associates are *Philanthus albopilosus, Microbembex hirsuta*, and *Bembix pruinosa*. Those species are, however, more characteristic of the central part of blowouts and windward slopes. *P. psyche* tends to establish territories and to nest on the periphery of such areas, where there is sparse vegetation. The species is active from late May into September and probably has overlapping generations during that period, at least in the southern parts of the range.

Our major study area was near Roggen, Weld County, Colorado, where we worked at intervals during June through September 1974–1983. Many nests and territories occurred here each year around the margin of a blowout in level to slightly sloping soil. The sparse vegetation consisted of blowout grass (*Redfieldia flexuosa*) and lanceleaf scurfpea (*Psoralea lanceolata*). O'Neill (1979, 1983a) has made a detailed study of male behavior at this site.

Briefer studies were made at four other sites, all ecologically similar. These were (1) LaJoya Wildlife Preserve, 32 km north of Socorro, Socorro County, New Mexico, May through July 1974–1976; (2) Great Sand Dunes National Monument, Alamosa County, Colorado, August 1964 and July 1978; (3) Monahans Sandhills State Park, Ward County, Texas, June 1974 and June 1976; and (4) San Rafael Desert, 65 km southwest of Green River, Utah, August 1981.

Females of this species make a series of nests, each lasting only a few days. Active nests tend to be somewhat clumped, although separated by 0.3 m or more; if nests are plotted over the season they appear very close to each other (Fig. 6-1). Males establish territories among the nests, much as in the much larger species *bicinctus*. Males and females become active each morning about 1000 and remain active until 1400–1600 hours.

Male Behavior

The following summary is based on research done near Roggen, Colorado, and reported by O'Neill (1979, 1983a); brief, confirming data from three other sites are considered later. On warm, clear days males appeared in midmorning and perched on vegetation (Fig. 6-2). Between 1030 and 1100 they began to occupy territorial perches in the center of patches of bare sand in areas of short, sparse vegetation (Fig. 6-1). The mean distance from a nest to the center of the nearest territory was only 35 cm ($N = 29$). Territorial perches occurred in the center of small bare patches. The mean size of the bare areas was 0.17 m^2 ($N = 12$).

Certain territories were occupied more frequently than others. Of 12 that were censused regularly, 3 were occupied at least 85% of the time of male activity, while 5 were occupied less than 30% of the time. Even on partially overcast days and during the latter part of the season, when few males were territorial, the most popular territories tended to be occupied. It was possible to record how long certain males that were watched continuously occupied certain territories on a given day. The mean duration of occupation was 80 minutes ($N = 25$). Males did not

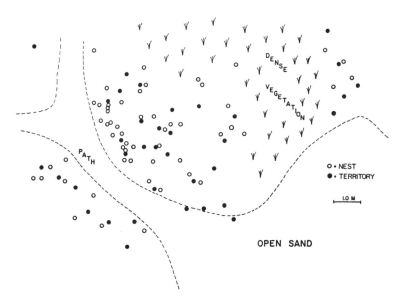

Figure 6-1. Map of main observation area for *Philanthus psyche*, Weld Co., Colo., showing location of vegetation, nests, and territories.

necessarily occupy the same territory on consecutive days, but they did tend to remain in the same general area. The maximum period over which a marked male was seen was 14 days, and the maximum number of territories a marked male was seen to occupy was nine. Males returning after the day on which they were marked established 52 territories; of these, 46 (88%) were within the area of grass in which the male was originally marked. Only six (12%) were outside this area, with only one farther than 5 m away.

Much of the male's time in the territory was occupied by interactions with other insects, particularly conspecifics. The major response to intruding males was a swirling flight consisting of two (or rarely three) males repeatedly circling one another in loops with a radius of 5–15 cm. These flights usually lasted from less than 1 second to approximately 10 seconds. They involved males holding adjacent territories or a territorial and a nonterritorial male. A second type of interaction involved a territorial male dropping from a height of 10–30 cm onto the back of a male perched on the ground. The strike was usually discontinued after a grapple of a second or less, and it seems likely that these males were pouncing on a potential mate and not initiating an aggressive interaction.

Figure 6-2. A male *Philanthus psyche* resting on a blade of grass before establishing a territory in late morning (Weld Co., Colo.).

Removal experiments were conducted to determine the relative body size of territorial versus nonterritorial males (floaters) (O'Neill, 1983a). The maximum number of replacers that occupied a single territory in one day was nine; the majority of original residents removed were replaced (51%, $N = 47$). The mean interval between removal and replacement was 7.8 minutes ($N = 44$). Obviously there was a large pool of nonterritorial males. Mean head width of original residents proved to be 2.11 mm (SD = 0.14, $N = 47$) and that of first replacers 1.95 mm (SD = 0.18, $N = 24$), a significant difference. There was no difference in head width between first and subsequent replacers. Replacers were smaller than original residents 79.5% of the time, larger 13.6% of the time, and approximately equal the remaining 6.8% of the time.

Territorial males pursued a variety of insects that flew across or near their territories. The distance they were pursued appeared to depend on their similarity to *psyche* in size, color, and flight pattern. Response was greatest to females without prey, which were pursued distances ranging from 1.4 to 4.0 m (mean = 2.45 m, $N = 13$). The pursuit of males averaged less (1–2 m; mean = 1.37 m, $N = 13$), and that of prey-laden females still less (averaging about 1 m). *Bembecinus nanus*, a wasp somewhat similar in length and color to *P. psyche* but more robust, elicited about the same intensity of response as prey-laden females. Other insects were merely approached or pursued a short distance, never more than 70 cm ($N = 110$) (Fig. 6-3; O'Neill, 1979).

Stems on the periphery of the territory were scent-marked in a manner similar to that used by other *Philanthus*. The average number of times a male scent-marked each minute was 1.50 ($N = 20$). Each scent marking was preceded by a weaving flight in which the male left his perch and flew back and forth perpendicular to the wind direction and facing into it (Fig. 6-4). Plants marked were usually upwind of the perch (they represented 88% of 206 scent markings recorded). When the wind direction changed, different plants were marked (Fig. 6-5). Wind velocity also influenced the frequency of scent marking. The mean wind velocity (measured every 15 seconds) and the wind velocity when the males initiated scent marking were recorded for six males (each was

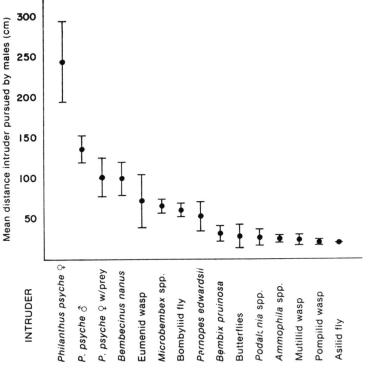

Figure 6-3. Intensity of response (distance pursued) of male *Philanthus psyche* for each species of intruder. Intruder species are arranged in order of decreasing similarity to a conspecific female. (From O'Neill, 1979, *Psyche*, courtesy of Cambridge Entomological Club.)

Species of the *politus* group   145

observed for 20 minutes). For these males, 76% of the scent markings were performed at wind velocities between 41 and 100 m per minute, although only 31.8% of the overall wind velocity readings were in this range. Evidently there is an optimal wind velocity for scent marking, possibly detected during the weaving flight that precedes each marking.

On clear days, sand surface temperature rose from 20–25° C early in the morning to over 50°C by 1300 hours. A maximum surface temperature of 60°C was recorded on a territory on the face of the dune. Males generally initiated territories each day when the surface temperature rose above 39–40°C. These temperatures, and even the slightly lower temperatures near the surface, presented males with potential thermal and hydric stresses. When 23 males were tethered at the sand surface across a variety of temperatures (40–59°C), a significant negative correlation was observed between surface temperature and the time it took the males to die from the heat. One male forced to remain on the surface at 59°C died 7 seconds later. Rather than abandon their territories and thereby avoid such stress, males adjust to the heat in two ways. First, as

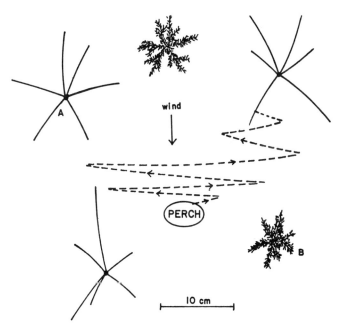

Figure 6-4. The weaving flight of scent-marking male *Philanthus psyche*. The dashed line indicates the flight path. A and B are plants bordering the territory. (From O'Neill, 1979, *Psyche*, courtesy of Cambridge Entomological Club.)

146 The Natural History and Behavior of North American Beewolves

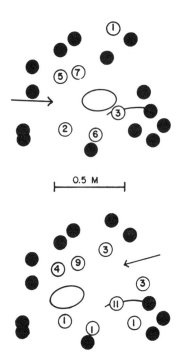

Figure 6-5. The effect of wind direction on position of scent-marked plants for one territory of *Philanthus psyche*. Arrows indicate wind direction. Solid circles are plants not scent-marked; hollow circles are marked plants. Numbers in hollow circles indicate the number of times plant was marked in 10 minutes (above) and in 15 minutes (below). The perch is indicated by an ellipse. (From O'Neill, 1979, *Psyche*, courtesy of Cambridge Entomological Club.)

the day progresses, they perch on the ground more briefly; when the surface temperature rises about 50°C, they rarely stay on the surface for longer than 1 or 2 seconds. Second, although territorial males perch on the surface of the sand 99% of the time early in the day, they switch primarily to plant perches as the surface temperature rises about 45°C. Such temperature-related behavioral adjustment not only reduces thermal stresses but may keep male body temperature within a range optimal for territorial activity (O'Neill and O'Neill, in press).

Three matings were seen in the nesting area and none elsewhere. Two of them were observed already in progress, the male and female coupled on a stem in the territorial area. The third copulation was initiated when a female flew upwind and landed on the edge of a territory. The resident male immediately pounced on her. They coupled at once and flew to a plant about 1 m away, where they remained still for about 4 minutes. At this point and again several minutes later a second male landed on the back of the female; as a result, the original pair changed perches once. They finally uncoupled after 8 minutes, 30 seconds.

Brief observations at LaJoya, New Mexico, and in the San Rafael

Desert of Utah confirmed the major aspects of male behavior. At La-Joya, males were spaced about in bare places among low vegetation, undertaking weaving flights and scent-marking stems upwind of the perches. Unlike those in the Roggen population, males here exhibited butting as well as swirling flights in their interactions. In the San Rafael Desert, two males had established territories 2 m apart. One of them, during the 10-minute period 1145–1155, scent-marked a grass clump and a small *Euphorbia* upwind of the perch 18 times in 13 bouts. The only interaction during this period involved the short pursuit of a tiphiid wasp that crossed the territory. Not far away, several males had established territories in bare places among vegetation along the bank of an intermittent stream. About half of them proved to *psyche* and the others *multimaculatus*; males of the two species appeared intermingled at this site.

Nesting Behavior

Active nests are typically spaced 0.3–3.0 m apart in places where there is sparse vegetation. Since the nests are quite shallow, they evidently benefit from the stabilizing effect of vegetation, but at the same time females require a small amount of space among the plants for digging. Nests are declinate and are temporary affairs, started in the morning (often 0900–1000 hours) and receiving a final closure the same day, in the late afternoon, or within 2–4 days at the most. The maximum number of days any nest was found to be active was 4, the maximum number of cells per nest four. Nests lasting but 1 day normally have a single cell.

Digging of the oblique burrow requires no more than 1–1.5 hours, during midmorning. The angle of this burrow with the slightly sloping surface varies from 20 to 35°. Prey is stored at or near the end of this burrow, and in the late afternoon or evening a more nearly vertical burrow is dug much more deeply into the sand; at the end of this burrow a cell is built, and the prey is removed into it (Figs. 6-6D, 6-6E; Table 6-1). This nearly vertical portion is closed off after oviposition and a similar one dug in a slightly different direction the following day (if the nest persists for more than 1 day). In multicellular nests the cells are separated by 7–16 cm.

Females dig the oblique burrow rapidly, throwing up a small mound of sand that when complete measures 7–12 cm long by 5–10 cm wide and about 0.5 cm deep in the center. Such leveling as occurs is performed during the later stages of digging or to a lesser extent imme-

148 The Natural History and Behavior of North American Beewolves

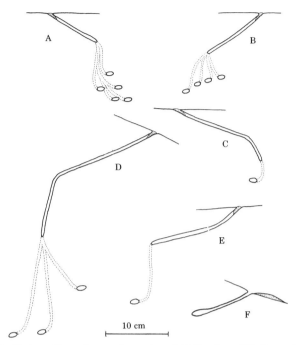

Figure 6-6. Nest profiles of several species of *Philanthus*. (A) *P. pacificus*, Jackson Hole, Wyo. (B) *P. politus*, Rensselaerville, N.Y. (C) *P. politus*, Bedford, Mass. (D) *P. psyche*, La Joya, N.M. (E) *P. psyche*, Roggen, Colo. (F) Sleeping burrow of a male *P. zebratus*, Jackson Hole, Wyo.

diately after the burrow has been completed. There is an unusual amount of variation in the extent of leveling, even among females nesting in close proximity. In a few cases the mound remains apparently wholly intact; at the other extreme it may be leveled virtually completely. More typically, only that part of the mound immediately in front of the entrance is flattened, the remainder intact or only slightly dispersed. The following notes on one female at LaJoya will demonstrate the usual pattern of digging, clearing, and leveling. The nest was started at about 0900 hours on 14 June 1976. Observations were continuous from 0935 to 1100.

    0936–0938:   Backed out 2–3 cm three times and went back in digging each time. When backing out, she did no digging; she simply moved out in a straight line, bobbing the abdomen up and down, then scraped soil as she moved to the entrance and into it.

Table 6-1. Nest dimensions (cm) of Philanthus psyche (mean and range of variation)

Locality	No. nests	Length of oblique burrow	Depth at end of oblique burrow	Depth of cells
Colorado: Weld Co.	14	15.3 (6-20)	5.7 (2-12)	21.5 (16-27)
Colorado: Alamosa Co.	2	10.5 (11-12)	6.5 (5-8)	20.0 (15-25)
New Mexico: Socorro Co.	7	20.5 (13-32)	12.0 (8-15)	33.5 (25-44)
Texas: Ward Co.	3	18.3 (15-23)	10.0 (6-13)	----

0940: Entrance blocked from inside (digging inside).
0944–0947: Backed out 1.0–2.5 cm eight times and went in digging (as described above).
0948: Entrance closed from inside (digging inside).
0950–0954: Backed out five times, 1–3 cm, remaining outside 8–10 seconds each time; went back in digging.
0955: Came out head first and scraped sand into entrance; then backed to middle of mound, worked forward with some zigzagging, and reentered at 0956 (first true leveling).
0956–1002: Backed out nine times, then reentered digging.
1003: Came out head first, closed, went to far end of mound, and leveled in a zigzag pattern to slightly beyond the entrance; did this twice and reentered at 1004.
1006: Began making accessory burrow on left side of entrance, 1 cm away.
1008: Accessory burrow complete, 1.5 cm deep; wasp reentered true burrow and closed from inside.
1013: Reappeared head first, closing behind her; flew in three loops about 10 cm high; returned to entrance; then made three wider and higher loops, 20 cm high with 1 m radius. Flew off, leaving entrance closed, accessory burrow open.

This wasp had not returned by 1100, but at 1206 she was seen returning with prey, remaining in the nest only 15 seconds. This nest was excavated at 1730 hours, at which time the female had begun the vertical part of the burrow. It lay at an 80° angle with the surface and reached a depth of 15 cm, 10 cm below the point where the prey had been stored. The mound remained largely intact, with a slightly flattened area at the entrance.

Other females were seen to reenter and close twice before starting the orientation flight. The majority of females omitted the accessory burrow. Only 1 of 14 freshly completed nests at Roggen had such a burrow. At LaJoya, 4 of 14 nests had an accessory burrow; these burrows were 0.3–1.5 cm long, and one of them persisted for 4 days.

Final closure of the nest involves complete filling of the oblique burrow and is usually made late in the afternoon. The female scrapes in sand from the side of the burrow and later from the periphery of the entrance, leaving a small star-shaped pattern in the sand, some 5 cm in diameter, with short, shallow grooves radiating from a small mound in the center. No nests at Roggen were found to persist for more than 1 day; hence the final closures were made the same day as the nest was initiated. At LaJoya, final closures were observed as many as 4 days after the nest had been initiated; such nests contained four cells.

Provisioning the Nest. The prey is diverse, consisting of various very small wasps and bees (Tables 6-2, 6-3). After storage at or near the bottom of the initial, oblique burrow, the prey is removed to a freshly prepared cell, 4 to 13 per cell (mean = 7.5, $N = 13$), the number depending at least in part on the size of the prey. In no case were more than 12 prey items found in storage in the oblique burrow. Thus it appears that as soon as enough prey has been taken, a cell is prepared, the prey removed to it, and an egg laid.

Natural Enemies

No parasite-avoidance behavior was noted on the part of provisioning females, and none of the cells excavated contained maggots ($N = 15$). Miltogrammine flies did not appear abundant in the nesting areas, but at LaJoya a *Senotainia rufiventris* was captured as it followed a prey-laden female.

As noted by O'Neill (1979) and by Gwynne and O'Neill (1980), territorial males are subject to predation by robber flies (Asilidae). These flies constituted almost 5% of the intruders in territories at Roggen and were often approached or pursued by males. Four attempted predations were observed, and two were successful. We also saw a female preyed on by a small wolf spider (Lycosidae) in the nesting area at Roggen.

Table 6-2. Wasps used as prey by Philanthus psyche

Species	Weld Co. Colorado		Alamosa Co. Colorado		Texas and New Mexico	
BRACONIDAE						
Ascogaster sp.						3 ♂
Chelonus sp.		2 ♂		2 ♂		
ICHNEUMONIDAE						
Anomalon sp.		2 ♂	3 ♀	1 ♂		1 ♂
Coccygomimus hesperus						1 ♂
Genus and species?	1 ♀		1 ♀		1 ♀	
CHRYSIDIDAE						
Elampus sp.					1 ♀	
Hedychridium fletcheri	1 ♀	1 ♂				2 ♂
Hedychrum violaceum		1 ♂				1 ♂
Parnopes fulvicornis					1 ♀	
Pseudolopyga sp.	1 ♀					
MUTILLIDAE						
Protophotopsis scudderi		1 ♂				
POMPILIDAE						
Anoplius cylindricus		1 ♂				
Tastiotenia festiva						1 ♂
SPHECIDAE						
Ammopsen masoni					1 ♀	
Ammopsen sp.	1 ♀					
Belomicrus cladothricis					2 ♀	1 ♂
Diodontus spp.	1 ♀		1 ♀		5 ♀	1 ♂
Entomognathus sp.					1 ♀	1 ♂
Larropsis sp.		2 ♂				
Lindenius sp.	3 ♀					1 ♂
Mimesa sp.	1 ♀	2 ♂				
Oxybelus abdominalis	1 ♀	3 ♂				
Oxybelus emarginatus	2 ♀	1 ♂				
Plenoculus sp.						5 ♂
Solierella sp.	1 ♀					
Tachysphex similis		2 ♂			1 ♀	2 ♂
Trichogorytes cockerelli					2 ♀	6 ♂
TOTALS 27 species	13 ♀	18 ♂	5 ♀	3 ♂	15 ♀	26 ♂

Philanthus tarsatus Smith

The species *P. tarsatus* is characteristic of the Great Plains, ranging from Nebraska to Texas. It is very similar to *psyche* in size and coloration and occurs in much the same habitat, areas of fine-grained sand with sparse vegetation. We have encountered the species in only one area, the sand dunes just northeast of Roggen, Weld County, Colorado. Males have been found at several sites here, at the edges of blowouts and along abandoned sandy roads. Unfortunately we have never found more than a few males, and we have found no nests at all, so we have no

Table 6-3. Bees used as prey by Philanthus psyche

Species	Weld Co. Colorado		Alamosa Co. Colorado		Texas and New Mexico	
COLLETIDAE						
Hylaeus cressoni mesillae					2 ♀	
ANDRENIDAE						
Nomadopsis sp.					1 ♀	
Perdita crotonis	1 ♀	2 ♂				
Perdita dolichocephala		11 ♂				
Perdita fallax		2 ♂	6 ♀	4 ♂		
Perdita hirsuta					3 ♀	6 ♂
Perdita ignota						2 ♂
Perdita maculigera	10 ♀	3 ♂				
Perdita punctosignata					1 ♀	
Perdita similis			2 ♀	3 ♂		
Perdita tridentata		5 ♂				
Perdita spp.	1 ♀		2 ♀	5 ♂		
Pseudopanurgus sp.				1 ♂		
HALICTIDAE						
Conanthalictus conanthi						1 ♂
Dialictus albohirtus			1 ♀			
Dialictus oleosus					2 ♀	
Dialictus pictus			2 ♀	2 ♂		
Dialictus pruinosiformis	1 ♀	4 ♂			2 ♀	
Dialictus pruinosus					1 ♀	
Dialictus tegulariformis	1 ♀				2 ♀	
Dialictus spp.	4 ♀	10 ♂			6 ♀	1 ♂
Dufourea sp.		1 ♂				
Evylaeus pectoraloides	1 ♀				1 ♀	1 ♂
Sphecodes spp.	4 ♀	2 ♂			1 ♀	1 ♂
MEGACHILIDAE						
Heriades variolosa					2 ♀	1 ♂
ANTHOPHORIDAE						
Neolarra vigilans	1 ♀					
Oreopasites sp.					1 ♀	
TOTALS 27 species	24 ♀	40 ♂	13 ♀	15 ♂	25 ♀	13 ♂

data on the relationship of male territories to nests. There are no published reports on the behavior of this species.

Male Behavior

Males have been seen at Roggen over the period 14 August to 2 September, and we judge this to be a species characteristic of late summer. They are territorial at much the same sites as *psyche* (we have sometimes found territories of the two species intermingled, although *tarsatus* tends to be active later in the day, 1200–1430 hours). Fortunately a minor color pattern difference made it possible to separate the two species in the field, and behavior was noticeably different, *tarsatus* males spending much time hovering downwind of tall plants. Perches were usually on plants 10–30 cm above the ground (Fig. 6-7). Although many spaces appeared to be suitable for occupation by males, in fact the few males present appeared to contest for only a few of them. Not uncommonly, two to four males appeared on or near a single territory, the nonresidents perching on plants away from the center of the territory. As the resident hovered and flew slowly from plant to plant he tended to encounter these intruders. Such encounters involved approach flights or swirling flights that caused intruders to leave, at least temporarily. We saw no bodily contact between males.

One male was seen to take over a territory at 1205 hours and to remain there until 1426. In this period of 2 hours and 21 minutes he scent-marked 237 times, always once per bout and always on a single tall plant upwind of the center of the territory. The frequency of marking did not diminish over this period; the rate varied from 14 to 22 marks per 10-minute period (mean = 16.9). On five occasions during this period the male defended the territory from the intrusions of other males, and on six occasions he took nectar from a *Croton* plant within the territory.

Figure 6-7. A male *Philanthus tarsatus* on a territorial perch on top of a plant, near Roggen, Colo.

On another occasion a male had established a territory downwind of an isolated clump of grass 0.5 m tall growing at the edge of a blowout. In the 10-minute period from 1321 to 1331 he scent-marked various blades of the grass clump 17 times, always once per bout. He did much hovering but occasionally perched on grass stems or short, dead weed stems; twice he perched on the ground briefly. During the 10-minute period he pursued a *Microbembex* a short distance, but there were no interactions with conspecific males. When this male was removed at 1334 he was replaced by another male in 1 minute. When this second male was removed, he was replaced within 1.5 minutes. Clearly there were floaters in the area ready to take over this territory, even though to us the spot looked no different from others that were never occupied. The resident male was found to have a head width of 2.48 mm. Both replacers had head widths of 2.25 mm and thus were appreciably smaller than the original occupant of the territory.

Philanthus serrulatae Dunning

P. serrulatae, another small species, ranges from South Dakota and Wyoming south through Texas and Arizona to central Mexico. *P. siouxensis* Mickel is a synonym (Ferguson, 1983). There are no published accounts of the biology of this species. We have studied it at two sites in north central Colorado. Site 1 was 8 km north of Fort Collins, Larimer County, where a small aggregation was studied 17–21 August 1978. Site 2 was located 18 km northeast of Nunn, Weld County, and studied 12 July–5 August 1982 and 7 July–4 August 1983. Individuals of both sexes were frequent visitors to flowers at both sites, at Site 1 principally *Eriogonum effusum*, at Site 2 principally *Melilotus alba*. At Site 2 members of both sexes also visited young *Helianthus* plants frequently, apparently taking nectar or honeydew from leaf axils. The species was originally described by Dunning (1898) from specimens taken on *Cleome serrulata*.

At Site 1, about 10 females nested in firm, compacted silt along a little-used dirt road between agricultural land and a reservoir. Nests were intermingled with those of *P. multimaculatus* and *Clypeadon laticinctus*. All *serrulatae* nests were within an area measuring 2 by 4 m, and most were grouped within 1 m^2, some with nest entrances only 15 cm apart. No observations on male behavior were made at this site. In subsequent years the site became overgrown with vegetation, and no *Philanthus* could be found nesting.

At Site 2 in 1982, 16 nests were located in an area 4 by 55 m along a sandy, little-used road through open prairie. Nests were in bare places in or adjacent to the tracks of the road. Surrounding vegetation consisted of grasses, forbs, and cacti. Along a similar track some 50 m to the east, 63 additional nests were counted on 24 July; they were distributed irregularly in bare places along the track for a distance of 50 m. Some nests were no more than 6 cm apart, but the majority were spaced 20–100 cm apart; in places where the soil was less sandy, gaps of 4–12 m contained no nests.

In 1983, the population appeared to have declined markedly, and only 18 nests could be located, all in an area of about 3 by 3 m where nests had been highly concentrated the previous year. Excavation of several nests revealed an abundance of old cells containing bee remains and cocoon fragments, suggesting that this site had been occupied for several seasons. In 1984 only one nest could be found at this site despite considerable searching, and in 1985 none at all.

Male Behavior

From 12 to 20 July 1982, several males at Site 2 were found to have established territories in the tracks of the road or in bare places immediately adjacent to the tracks. Territories were approximately 0.5 m in diameter; in no cases were there nests within territories, although some were within 2 m of a territory. Interactions between territorial males and intruding males consisted of swirling flights, butting, and grappling. Males scent-marked grass stems, usually upwind of the perch, by walking down the stems 8–15 mm; sometimes this downward movement was preceded by a short upward movement of only 1–2 cm. Perches were on grass stems or low weeds, 3–30 cm above the ground. Three males, watched for a total of 63 minutes between 1200 and 1400 on 2 days in 1982, scent-marked 142 times, each bout consisting of 1–4 scent-marks on different stems and followed by a return to the perch. Males changed from perching primarily on the ground to perching primarily on plants as the surface temperature rose.

In 1982, no territorial males were observed after 20 July. This fact, plus the fact that we saw no more than four territorial males at any one time (despite the large number of nests) leads us to believe that most territorial behavior had ceased by the time we began our observations. In 1983, we observed only one territorial male, on 7 July, a date on which we were unable to find any nests. However, several nests had appeared in the immediate vicinity of this territorial site by 15 July.

Males were collected as late as 4 August, but despite effort we were unable to find any others exhibiting territorial behavior or otherwise interacting with conspecifics.

Nesting Behavior

Nests at both sites were easy to locate, because mounds at nest entrances were never leveled and often persisted even beyond the duration of activity at the nest. Fresh mounds measured 5–8 cm long by 3.5–6 cm wide. Many nests had an accessory burrow within about 1 cm of the true nest entrance and approximately at a right angle with it. These accessory burrows were left open and tended to persist for several days. On 30 July five fresh nests were marked at Site 2, three of which had accessory burrows. On 4 August these same three nests had accessory burrows of identical form and depth, even though they were no longer active; the other two still had none. On the other hand, two nests marked on 27 July as lacking accessory burrows both had them on 30 July; both were still active. Thus it appears that such burrows are usually dug when the nest is prepared but are occasionally added later.

Of five nests studied at Site 1, three had one accessory burrow each, the others none. At Site 2 in 1982, of 21 fresh nests studied carefully, 1 had two accessory burrows, 11 had one, and 8 had none. A broader survey of 63 nests found on 24 July revealed that 27 had one accessory burrow each and 36 had none; not all of these nests were freshly prepared. Of 10 fresh nests studied carefully in 1983 at this site, 6 had one accessory burrow each and 4 had none. The single nest found in 1984 had two accessory burrows. Depth of the accessory burrows at Site 1 varied from 0.2 to 2.0 cm (mean = 0.8, $N = 5$); at Site 2 depth varied from 0.3 to 7.0 cm (mean = 2.6, $N = 16$).

Nest Structure. Nest structure was very similar at the two sites. The burrow entered the soil at a 35–50° angle with the surface, then, at a distance of 10–15 cm from the entrance, angled downward sharply, forming an angle of about 70° with the surface. Cells were constructed near the bottom of the steep portion of the burrow, at a distance of 17–35 cm from the entrance (Fig. 7-1C). Cells were spaced 5–8 cm apart and were subspherical, measuring about 5–8 mm. At Site 1, cells varied in depth from 11 to 20 cm (mean = 14.6, $N = 6$). At Site 2 in 1982, cells varied in depth from 6 to 12 cm (mean = 9.6, $N = 8$); at this site in 1983, cells varied in depth from 17 to 23 cm (mean = 19.8, $N = 5$). The great difference between the two years may be related to the fact that 1983 was a much wetter year, and the soil was consequently less dry and compacted. The maximum number of cells found per nest was three at both sites. That may approximate the maximum number of cells per nest. Several nests marked on 30 July were inactive and had final closures 5 days later.

At both sites it was noted that the female, having completed a nest, made a closure by scraping in soil from the periphery of the hole; then she scraped it open, reentered briefly, then emerged and repeated the closure. This procedure was sometimes repeated a second time before she made an orientation flight consisting of several irregular loops, gradually ascending to about 0.5 m before taking off. An outer closure was maintained at all times except during the brief period the female was bringing prey to the nest.

Provisioning the Nest. The prey consists of various small bees and wasps, the bee genera *Perdita* and *Dialictus* being most commonly used (Table 6-4). Prey was, as usual, allowed to collect in the bottom of the oblique portion of the burrow before being carried to a cell, 8 bees being the maximum found in storage. The number of prey provided per cell varied greatly depending on the size of the prey. At Site 1, from 5 to 10

Table 6-4. Prey records for Philanthus serrulatae

Species	Larimer Co. Colorado	Weld Co. Colorado
MUTILLIDAE		
Pseudomethoca frigida		1 ♂
SPHECIDAE		
Dienoplus sp.	1 ♂	
Diodontus sp.	1 ♂	
ANDRENIDAE		
Calliopsis andreniformis	4 ♂	
Hypomacrotera callops	3 ♀ 1 ♂	
Perdita spp.	3 ♀ 2 ♂	8 ♀ 12 ♂
HALICTIDAE		
Dialictus pictus		2 ♀ 2 ♂
Dialictus pruinosiformis	3 ♀	16 ♀ 18 ♂
Dialictus pruinosus		1 ♀ 1 ♂
Dialictus tegulariformis	1 ♀	
Dialictus spp.	2 ♀ 6 ♂	
Evylaeus pectoraloides	1 ♂	1 ♂
Halictus confusus		3 ♀ 2 ♂
Pseudopanurgus sp.		1 ♀
Sphecodes spp.	1 ♂	1 ♀ 14 ♂
TOTALS 15 species (3 wasps, 12 bees)	12 ♀ 17 ♂	32 ♀ 51 ♂

were found per cell (mean = 7.6, $N = 5$), at Site 2 from 5 to 17 per cell (mean = 10.2, $N = 6$). No parasites were found in cells or in the vicinity of provisioning females.

Philanthus politus Say

P. politus occurs throughout much of the eastern half of the United States as well as southeastern Canada. Evans and C. Lin (1959) reported on several nests from central New York. The report in this section is based on several nests found in eastern Massachusetts in 1969 and on several found near Rensselaerville, New York, in 1970. Brief observations were made on male behavior in Bethany, Connecticut, in 1980. *P. politus* nests in fine-grained to moderately coarse sand in man-made sandpits or on the periphery of dunes. It is bivoltine, at least in New York. Near Rensselaerville, we found numerous nests between 29 June and 19 July, then after a hiatus a large number of new nests appeared during the first 2 weeks of August. Males have been observed in June on the flowers of *Aralia hispida* and *Ceanothus americanus* and in August on *Daucus carota*. Both sexes have been collected on *Achillea millefolium* and on species of *Solidago*.

It is not uncommon to find isolated nests of this species or to find nests widely scattered among nests of *Bembix americana spinolae* or other digger wasps. On the other hand, nests are sometimes aggregated. Near Rensselaerville, we found about 50 nests in late June and early July, all in an area measuring about 3 by 6 m, with some entrances only 10–30 cm apart. In this same area, on 5 August, we counted about 100 nests of the second generation. Although *P. solivagus* and *P. lepidus* occurred in this same site, these species had scarcely begun nesting before the disappearance of *politus*. There was some intermingling of nests of *politus* and *gibbosus* during July and early August, but the latter species usually nested in coarser and firmer soil and frequently occupied slopes. All nests of *politus* were in flat or slightly sloping soil.

Male Behavior

Unfortunately our observations on males were made in the absence of knowledge of nests, so we have no data on the spatial relationships of male territories and female nests. Our observations were made in a large, man-made gravel pit in Bethany, Connecticut, over 2 days in early August. Several males were found to have established territories in a thinly vegetated area of pale, very friable sand. Each male perched downwind of an isolated bushclover plant (*Lespedeza capitata*) 0.4–0.8 m tall. None of the territories were closer together than 2 m. Detailed notes were made on one territory that was occupied for a second day. The occupant perched on the ground or low plants downwind of and within 50 cm of a bushclover. All scent markings were on this same plant, though on various branches; they consisted of the usual up-down pattern. Interactions with other males consisted of swirling flights and grappling. The male that occupied this territory was unmarked, but it was a large male, and we feel confident that he successfully defended this territory during the period of observation.

This male was seen to initiate territoriality at 1102. In the first 10 minutes he scent-marked 14 times in six bouts, once marking four different stems of the bushclover plant in a bout. During this period he engaged in four swirling flights with intruding males and one grapple to the ground. During the next 10 minutes he scent-marked 25 times in 21 bouts, engaged in eight swirling flights and two grapples with other males, and pursued a small butterfly and an asilid fly. Approach to scent marking was very much as described for *psyche*. Observations were terminated after this 20-minute period.

Allan Hook (personal communication, 1983) has observed male territoriality in Ocean Township and Tinton Falls, New Jersey, in late July and early August. In one instance males were situated in several locations around a field of about 300 m^2.

Nesting Behavior

Nests are commonly begun in the midmorning hours. The female requires only 2–3 hours for completion of the oblique burrow (Fig. 6–8). She throws sand back beneath her body in the usual manner, forming an elongate mound. Construction was described as follows by Evans and C. Lin (1959, p. 117):

> At times the earth is allowed to plug up the entrance, and the wasp comes out periodically and clears it away. Interspersed with digging are movements of leveling the mound of sand which accumulates; when the burrow is completed the wasp continues to level for as long as 30 minutes, so that no trace of the mound remains. In leveling, the wasp backs out in a straight line from the entrance, then works forward in a straight line or with a slight turning from side to side, kicking sand vigorously with the front legs. Upon reaching the nest entrance, she sometimes enters, to reappear in a moment head first, turn around, and again back out from

Figure 6-8. A female *Philanthus politus* digging at her nest entrance, Bedford, Mass. (From Howard Ensign Evans, *Wasp Farm*, 1963, courtesy of Doubleday and Co., Inc.)

the entrance. At other times she works past the nest entrance kicking sand, takes flight briefly, lands at the far end of the mound, then resumes kicking sand as she proceeds toward the entrance.

Leveling movements interspersed with digging are performed while the entrance is open; leveling performed after completion of the burrow occurs after an initial closure has been made. As leveling approaches completion, or following leveling, the wasp may reenter the nest one or more times, restoring the closure each time she leaves. The orientation flight consists of a flight of gradually increasing height from the entrance. We found no evidence of accessory burrows in this species.

Although the mounds outside newly completed burrows normally undergo a complete or nearly complete leveling, additional sand may be thrown up at later times, and the mound it forms is not usually leveled. Thus it is not unusual to see small mounds, up to 8 cm long by 5 cm wide, at nests that are being provisioned.

Nest Structure. Nests are declinate, as in other species of this group (Fig. 6-6B, 6-6C). The oblique burrow forms an angle of about 20–40° with the surface. Bees are stored at or near the end of this burrow, and when several have accumulated the burrow is extended downward at a much steeper angle, 60–90°. At the end of this burrow the first cell is prepared and the bees removed to it. After oviposition this steepened portion is closed off, at least toward the bottom; and after further accumulation of bees (usually the following day) another cell is prepared. Most of the nests we excavated had only two or three cells; several of these nests had received a final closure and hence had the definitive number of cells. Near Rensselaerville, we found one nest with four cells, one with five, and one with six. The five-celled nest is known to have been provisioned over at least 4 days, the six-celled nest over at least 3 (probably more). All marked nests became inactive after only 2–5 days, and it seems probable that females prepare about one cell per day, then after a few days make a permanent closure and prepare a new nest.

Cells measure 12–15 mm across and 8–10 mm high; they are subspherical or somewhat flattened on the bottom. Cells in any one nest tend to be spaced 3–8 cm apart; they often differ considerably in depth (e.g., 11–16 cm in one nest). Nest dimensions are summarized in Table 6-5.

Provisioning the Nest. Females carrying prey fly into the nesting area only 20–40 cm high and often land on vegetation or on the ground before proceeding to the nest. At the entrance, the closure is removed with several scrapes of the front legs. Occasionally the bee becomes lodged in the entrance; when this happens it is released and a moment later pulled in from the inside. Females often provision quite rapidly. One individual in Ithaca, New York, watched for 1 hour in the afternoon, brought in 9 bees at intervals of 2–13 minutes, each time remaining within the nest only 15–40 seconds (Evans and C. Lin, 1959). On

Table 6-5. Nest dimensions (cm) of Philanthus politus (mean and range of variation)

Locality	No. nests	Length of oblique burrow	Depth at end of oblique burrow	Depth of cells
New York: Ithaca	6	9.8 (8-14)	5.3 (4-9)	11.2 (7-16)
New York: Rensselaerville	6	8.6 (4-10)	3.6 (1-8)	11.8 (8-17)
Massachusetts: Concord and Andover	2	12.0 (8-16)	5.0 (4-6)	14.3 (13-16)

rare occasions the female leaves the entrance open upon leaving, but normally it is closed with a few strokes of the front legs. At night and during periods of inactivity it is always closed.

Prey consists of small bees, with a single record for a small wasp (Table 6-6). Of 379 prey records from five localities, 94% are Halictidae and 78% the halictid genus *Dialictus*; 78% are females. Nests dug at various stages of provisioning had from 1 to 9 bees in storage. At Ithaca, the number of bees per cell varied from 9 to 18 (mean = 13, $N = 18$). The number of bees per cell near Rensselaerville was smaller. During the first generation the number varied from 5 to 8 per cell (mean = 6.6, $N = 13$); during the second generation (after 5 August) the number varied from 7 to 11 per cell (mean = 8.1, $N = 10$).

Final Closure. We observed two final closures near Rensselaerville, both in the midmorning hours. In both cases the female came out head first and bit soil from the side and especially the front of the hole. She scraped the loosened soil into the hole with her front legs and turned around and went in head first. She then reappeared head first and pounded the soil in place with light, quick blows with the tip of the abdomen. Following this firming of the soil she came out and repeated the entire process. After a few minutes the burrow was completely full. She then began to scrape sand toward the filled hole from all sides, forming a series of radiating lines up to 5 cm long. Fully 10 minutes were employed making these lines, after which time the wasp hovered briefly over the site and flew off.

Natural Enemies

No parasitic flies were noted at Ithaca, but at Granby Center, New York, prey-laden females were often trailed by the miltogrammine fly *Senotainia vigilans* (Evans and C. Lin, 1959). Near Rensselaerville, New York, we saw females followed by *S. trilineata* on numerous occasions, and on two different occasions flies of this species were reared from nest cells. We also reared one specimen of the hole-searching miltogrammine fly *Phrosinella fulvicornis*. Of 36 cells excavated near Rensselaerville, 9 either contained maggots or appeared to have had the contents destroyed by maggots. At Concord, Massachusetts, five *S. trilineata* adults were reared from a single nest. We noted no evasive flights on the part of *politus* females, and it appears that they are not highly successful in avoiding the attacks of miltogrammines.

Near Rensselaerville, several females were seen to be struck in the

Table 6-6. Prey records for Philanthus politus

Species	Central New York[1]		Eastern New York[2]		Eastern Massachusetts[3]	
SPHECIDAE (CRABRONINAE)						
Oxybelus uniglumis				1 ♂		
COLLETIDAE (HYLAEINAE)						
Hylaeus affinis	1 ♀	2 ♂				
Hylaeus ellipticus			3 ♀			
Hylaeus mesillae cressoni			2 ♀			
Hylaeus modestus	4 ♀	1 ♂	2 ♀		1 ♀	
Hylaeus verticalis	1 ♀					
ANDRENIDAE (PANURGINAE)						
Calliopsis andreniformis	1 ♀		1 ♀			
Perdita octomaculata			1 ♀			
Pterosarus andrenoides			1 ♀			
HALICTIDAE (HALICTINAE)						
Augochlora pura	1 ♀					
Augochlorella striata	9 ♀	1 ♂	5 ♀	2 ♂		
Dialictus albipennis	2 ♀		1 ♀			
Dialictus alternatus		1 ♂				
Dialictus cressonii		1 ♂	3 ♀			
Dialictus disabanci			2 ♀			
Dialictus heterognathus			7 ♀		3 ♀	
Dialictus imitatus	68 ♀	42 ♂	36 ♀			
Dialictus laevissimus			6 ♀			
Dialictus lineatulus	2 ♀					
Dialictus nymphaearum	1 ♀		1 ♀			
Dialictus oblongus			7 ♀			
Dialictus philanthanus		5 ♂				
Dialictus pilosus	3 ♀		5 ♀	1 ♂		
Dialictus rohweri			59 ♀	1 ♂	1 ♀	
Dialictus solidaginis		2 ♂				
Dialictus tegularis				1 ♂		
Dialictus zephyrus	2 ♀		1 ♀	1 ♂		
Dialictus spp.	5 ♀	1 ♂	21 ♀		1 ♀	1 ♂
Evylaeus cinctipes			7 ♀		1 ♀	
Evylaeus divergens			2 ♀			
Evylaeus foxii					1 ♀	
Evylaeus sp.		1 ♂		2 ♂		
Halictus confusus	1 ♀	5 ♂	7 ♀	1 ♂		
Paralictus asteris	5 ♀	9 ♂				
Sphecodes spp.		1 ♂		2 ♂		
TOTALS 35 species (1 wasp, 34 bees)	106 ♀	72 ♂	180 ♀	12 ♂	8 ♀	1 ♂

1. Ithaca and Granby Center.
2. Rensselaerville.
3. Concord and Andover.

air by conopid flies, *Zodion americanum*. Members of this genus are normally parasites of bees, and it is uncertain whether the larvae normally develop successfully in adult *Philanthus*.

Philanthus parkeri Ferguson

A recently described species, *P. parkeri* (Ferguson, 1983), is the smallest North American *Philanthus*, varying in length from 5 to 8 mm. It was first discovered in the San Rafael Desert, Emery County, Utah, but is now known to range from southern Idaho to northern Arizona, with a few records from east of the Rocky Mountains (Nebraska, New Mexico). Most specimens have been taken in August and September. Stubblefield (1985) has reported that males are territorial but leave the territory promptly when approached by a predator such as an asilid fly or a female *P. basilaris*. He has recorded the following prey: *Pluto* sp. (Sphecidae), *Dialictus* sp. (Halictidae), and five species of *Perdita* (Andrenidae).

Chapter 7
Other North American Species of *Philanthus*

The five species considered here form a diverse lot structurally. In no case has the behavior been thoroughly studied, although some notes are available on all five. Bohart and Grissell assigned these species to four separate species-groups. One species, *albopilosus*, lacks clypeal brushes and ventral abdominal hairs, and in the others the abdominal hairs are of variable development, often reduced. We consider these species in the following order: *ventilabris, lepidus, bilunatus, solivagus,* and *albopilosus.*

Philanthus ventilabris Fabricius

P. ventilabris is one of the most widely distributed North American *Philanthus* species, ranging transcontinentally from southern Canada to Florida and to central Mexico. There are several reports on its biology, but all are brief (Peckham and Peckham, 1905; Rau and Rau, 1918; Evans and C. Lin, 1959; Alcock, 1975a, 1975c). Although we have collected individuals many times, we are able to report on only two nests. Evidently this is a relatively solitary species, showing little tendency to form nesting aggregations. Both males and females are frequent visitors to flowers; we have taken them on *Tamarix* and *Eriogonum;* the Raus (1918) report visitations to *Solidago,* and Krombein (1936) to *Solidago, Achillea,* and *Daucus carota.* The species does not appear at all restricted as to habitat, occurring both in semidesert and in more humid areas; nests are dug in quite firm soil, so there is no requirement for a sandy substrate.

Male Behavior

The Raus found a male apparently spending the night in a tiger beetle burrow, presumably abandoned. In Arizona, Alcock discovered two males that shared a burrow with a female. He also saw two males occupying a tall dried weed within 10 m of two known nests. They were seen to drag their abdomens on the weed and to fly out periodically; while perched they assumed an "alert" posture, with antennae extended forward and front legs elevated. According to Alcock (1975c, p. 541), "the hover-approach of a visitor, aerial pursuit, and the slow vertical flight with one male trailing another are part of this species' behavioral repertory." He found no perch to be occupied for more than an hour, although he found perches to be occupied intermittently over several weeks (7 June–15 July 1974).

Allan Hook (personal communication, 1983) observed males scent-marking on salt cedar (*Tamarix*) seedlings about 1 m tall in Big Bend National Park, Texas, in June 1980. Others were marking seep willow (*Baccharis*). Territories were located in gravel beds along the Rio Grande River only 0.2 m above the water level.

Nesting Behavior

We observed several individuals of both sexes along a dirt road about 8 km north of Fort Collins, Colorado, 20–30 August 1978. One female was seen digging in the hard-packed clay-silt of the road on 21 August. The somewhat fan-shaped mound she produced measured 10 cm long by 7 cm wide at its widest and had several shallow grooves leading from the entrance, made by the female as she backed out periodically, scraping soil. After completion of this digging, she dug an accessory burrow 2.5 cm from the entrance and at about a right angle to it. This accessory burrow initially measured 3 cm; later it became partially filled with soil, but it persisted until the nest was excavated 9 days later. The true nest entrance was, however, kept filled with soil whenever the female was away or was within the nest for a long period.

This nest was surprisingly shallow. The burrow entered the soil at an angle of about 30°, then, after 15 cm, turned to the right and dipped sharply downward for another 7 cm, reaching a final depth of 15 cm. There were seven cells, at depths varying from 14 to 18 cm. This nest had been started 9 days earlier; thus the female had built about one cell a day. The cells were distributed on each side of the burrow in a somewhat irregular pattern. Two contained eggs, two larvae, and three cocoons; in general, the cells with cocoons were somewhat farther from the entrance than the newer cells. Cells measured 12 by 20 mm and were separated by 4–6 cm of soil.

A second nest was found in a ridge of rather hard-packed loamy soil along an irrigation ditch in Fort Collins, Colorado, 13 July 1981. The mound of soil at the entrance had been eroded away when the nest was found. The burrow entered the gentle slope of the ridge at an angle of about 50° with the slope and terminated at a depth of 18 cm, 30 cm from the entrance (Fig. 7-1A). One bee was stored at the bottom of the burrow, and the female arrived with another bee while we were excavating the nest. There were eight cells, at vertical depths of 16–24 cm, 20–36 cm from the entrance. The cells measured about 9 by 14 mm and were arranged along the burrow, separated from the burrow by 2–3 cm. The cells closest to the entrance contained cocoons, whereas those deepest in the soil contained eggs.

The nest reported by Evans and Lin from near Austin, Texas, had a single cell only 15 cm deep. In contrast, the Raus (1918, p. 117) remarked that the nest they excavated in Missouri was "one of the longest wasp tunnels that we have seen" (they gave no details). Two nests excavated by Alcock (1975a) in Arizona were also long and deep; the two were 10 cm apart in "compact clayey sand" (p. 164). The burrows measured 105 and 123 cm and reached depths of 52 and 60 cm. Both started out at a low angle with the surface, then dipped downward sharply and

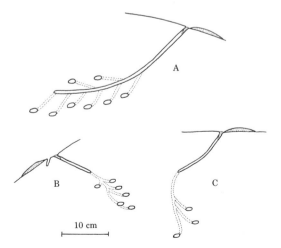

Figure 7-1. Nest profiles of three species of *Philanthus*. (A) *P. ventilabris*, Larimer Co., Colo. (B) *P. bilunatus*, Larimer Co., Colo. (C) *P. serrulatae*, Weld Co., Colo.

had one or more loops. Cells were constructed at the ends of short cell burrows and had a mean depth of 41 cm ($N = 14$). One nest had two "double cells," that is, one just beyond the other off the same cell burrow. Both nests had accessory burrows and unleveled mounds.

According to Alcock (1975a), one of the females he studied occupied a nest that had evidently "been built the preceding year from which she and several other individuals may have emerged earlier in the season" (p. 164). This female shared her burrow with two or three males. Thus there is evidence that this species resembles *gibbosus* with respect to reoccupation and sharing of nests.

Allan Hook (personal communication, 1983) found a female and two males in a nest he excavated in Big Bend National Park, Texas, in July 1980. This nest was found to contain 16 cells, mostly old cells with prey remains, suggesting reuse of the burrow by a second generation. He confirmed the presence of accessory burrows in nests at this site (one nest had two).

Although there is some ambiguity concerning nest structure in this species, most reports suggest a serial pattern of cells in a proclinate nest.

Provisioning the Nest. Alcock (1975a) observed that prey-laden females, if disturbed, would fly to a plant several meters from the nest and wait there, sometimes up to 5 minutes, before returning to the nest "in a rapid, sinuous flight about 50 cm above the ground. Only when a few cm from the nest did the female slow her approach and drop down the remaining distance (5 to 10 cm) to the entrance, which she opened immediately" (p. 164). He reported an average of 16 minutes between trips with prey ($N = 10$).

All records indicate that a closure is maintained while the female is away from the nest. We found 9–11 prey items in three cells of the first nest we excavated, while the second contained 16–21 prey items in the two cells that were fully provisioned. Prey records are summarized in Table 7-1. In addition, Evans and C. Lin found bees of two species in a nest excavated in Texas: *Perdita* sp. (Andrenidae) and *Halictus ligatus* (Halictidae). The Peckhams reported that the species uses "bees of several genera."

Natural Enemies

One of the eight cells in the nest we excavated in Fort Collins contained maggots, and a second appeared to have had the contents de-

Table 7-1. Prey records for Philanthus ventilabris

Species	Larimer Co. Colorado		Maricopa Co. Arizona[1]
TIPHIIDAE			
Genus and species?			x
VESPIDAE			
Stenodynerus sp.	1 ♀		
SPHECIDAE			
Oxybelus pitanta		1 ♂	
COLLETIDAE			
Colletes phaceliae	1 ♀	7 ♂	
ANDRENIDAE			
Calliopsis andreniformis	6 ♀		
Perdita albipennis			x
HALICTIDAE			
Agapostemon angelicus		3 ♂	
Agapostemon virescens		1 ♂	
Dialictus microlepoides			x
Dialictus occidentalis	1 ♀		
Dialictus pruinosiformis	4 ♀	2 ♂	
Dialictus zephyrus	1 ♀		
Dialictus sp.	2 ♀		
Evylaeus sp.		2 ♂	
Halictus confusus	10 ♀		
Halictus ligatus			x
Lasioglossum sisymbrii			x
Nomia nevadensis arizonensis			x
Sphecodes spp.	2 ♀	3 ♂	
Sphecodogastra oenotherae	1 ♀		
TOTALS 20 species (3 wasps, 17 bees)	29 ♀	19 ♂	

1. Data from Alcock (1975a). The number of specimens was not stated.

stroyed by maggots. The maggots pupated within a few days and three weeks later gave rise to adult flies, *Metopia argyrocephala*.

Philanthus lepidus Cresson

P. lepidus is a species of wide but somewhat localized distribution in the eastern half of the United States. Typically it occurs in areas of sparsely vegetated, fine-grained sand, either on the periphery of dunes or in man-made excavations. Evans (1964b) reported on the nesting behavior in central New York and eastern Massachusetts. The present report covers 10 additional field notes from Massachusetts and 14 field notes made near Rensselaerville, New York, in the late summer of 1970.

P. lepidus is a distinctly autumnal species, making its appearance in the first week of August and nesting through September (the latest observed nesting date was 26 September, near Rensselaerville). It is one of the only digger wasps of any group to be found nesting so late at this latitude. The females are sometimes active under cloudy and cool conditions. We reported several "digging sluggishly" at 16°C after several nights of frost (Evans, 1964b). We once observed a female bringing in prey in a light rain. The commonest associate of *lepidus* is *solivagus*, but the latter species more often nests in banks and has largely finished nesting by mid-September. Nests of *lepidus* usually are made in flat or nearly flat sand, although we have occasionally found nests in 20–30° slopes.

Both male and female *lepidus* commonly visit flowers of *Solidago* in late summer. Males are also sometimes seen in the nesting areas, landing on the sand here and there with their antennae extended rigidly forward and now and then pursuing females. In the late afternoon they are often seen entering holes in the sand. Males have been taken as late as 14 September in eastern Massachusetts. We have made no observations on male reproductive behavior.

Nesting Behavior

Nests tend to be clumped in bare places, as many as 20–30 being grouped so that the entrances are only 10–50 cm apart. As many as 100 nests were located in several clusters in flat sand just north of New Haven, Connecticut, but it is unusual to find that many nests at one site. At Lexington, Massachusetts, females nested in the same site for at least 8 years (1962–1969) until the site was devastated by motorcycles.

"From 3 to 6 hours are required to complete the burrow. The sand is allowed to plug the entrance, and from time to time the wasp comes out and clears it away, sweeping it into a broad mound in front of the opening. Mounds of completed nests measure from 8 to 13 cm in length by 3 to 8 cm in width and 0.5 to 1.5 cm in depth. No true leveling movements occur at any time, but mounds may weather away after several days, particularly if there has been a heavy rain or strong wind" (Evans, 1964b, p. 144).

The most characteristic feature of the nests of this species is the presence of one or more accessory burrows dug at an angle to the true burrow or sometimes into the mound. These burrows are left open, whereas the true burrow is closed at all times except when the female is

Other North American Species of *Philanthus* 171

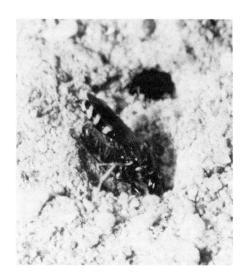

Figure 7-2. A female *Philanthus lepidus* entering her nest with a bee. An open accessory burrow is visible.

actually entering or leaving; this arrangement presents a confusing picture to the observer (Fig. 7-2). Although we have occasionally seen a nest entrance without an accessory burrow, the same nest a day or two later may be found to have one or more. Females are sometimes seen entering or even resting briefly in accessory burrows. These burrows are dug after the initial closure and thus do not serve as quarries for soil for closure.

Data on the incidence and depth of accessory burrows are presented in Table 7-2. It should be pointed out that we observed many more nests with accessory burrows than the 36 on which we took notes. The following observations, made near Rensselaerville, New York, demonstrate the persistence of these burrows:

Table 7-2. Accessory burrows of Philanthus lepidus

Locality	No. accessory burrows per nest					Length of accessory burrows
	1	2	3	4	5	
New York: Granby Center	3	1	0	0	0	6.5-9.0 cm
New York: Rensselaerville	6	8	1	0	0	0.5-10.0 cm
Massachusetts: Bedford and Lexington	7	4	3	2	1	0.5-10.0 cm

(1) On 21 August, in the morning, female dug an accessory burrow on the right side, leaving it after it was 2 cm long. On 22, 25, and 28 August the accessory burrow remained the same. On 29 August she was observed digging in the accessory burrow again.

(2) Female finished nest at 1105 on 22 August, closed burrow, and began an accessory burrow on right side of entrance. A few moments later she reentered true burrow, came out and closed, and began an accessory burrow on the left side of the entrance. She persisted here for 15 minutes, the resulting burrow being 7 cm long. She then went back into the true burrow, came out and closed after a few seconds, then went back to the accessory burrow on the right side and lengthened it from 0.5 to 1.0 cm. She reentered the true burrow again for a few seconds, came out, and flew off after a brief orientation flight.

(3) Female was seen digging new nest at 1030 on 28 August. On the following day at 1145 she was seen making an accessory burrow on the left side and then on the right side, both very short. She then reentered the true burrow, came out and closed, and resumed work on the two accessory burrows alternately until 1202, when she reentered the true burrow and closed from the inside. On 2 September the accessory burrow on the left side had been lengthened from 1 to 2 cm, but there was none on the right side. On 6 September the nest was still active but lacked accessory burrows.

Following completion of a new burrow, and sometimes at an older nest that has been disturbed, the female makes an orientation flight consisting of several loops only 10–30 cm above the ground. These loops tend to increase gradually in diameter. One such flight lasted fully 2 minutes and consisted of more than 30 loops or passes over the entrance.

Nest Structure. Nests are proclinate with a somewhat irregular pattern of cells. The true burrow enters the soil at an angle of only 15–30°, then after 6–15 cm descends sharply, forming an angle of 40–90° with the horizontal. Burrow diameter is only 5 mm, although the initial 6–15 cm may be somewhat wider, sometimes up to 12 mm. The descent of the burrow is usually regular and without undulations, although there may be one or more lateral curves. Cells are constructed individually at the ends of cell burrows 5–15 cm long. Data on burrow length and cell depth are presented in Table 7-3. Within any one nest the cells are at approximately the same depth; for example, in one nest all 11 cells were between 27 and 33 cm deep, and in another all 17 were between 25 and 33 cm deep (both nests were near Rensselaerville). These two nests, dug late in the season, 6 and 26 September, had the most cells found in any

Table 7-3. Nest dimensions (cm) of Philanthus lepidus (mean and range of variation)

Locality	No. nests	Burrow length	Cell depth
New York: Granby Center	1	48	34 (33-35)
New York: Rensselaerville	7	41 (35-48)	28 (17-33)
Massachusetts: Bedford and Lexington	3	28 (24-32)	25 (24-26)

nests. Another nest, initiated on 21 August, was still active when excavated 16 days later. However, a number of other nests marked as they were being started became inactive after 8–10 days, and it was common to see new nests started late in the season. Thus it appears that females do not usually maintain a single nest for a lifetime but may make two or more successive nests.

In general, the first cells are closest to the entrance and later cells deeper in the soil, although the pattern is by no means regular. Cells are of two sizes, larger cells measuring about 9 by 18 mm and stocked with 9–11 bees and smaller cells about 8 by 13 mm and stocked with 5–8 bees. It is assumed that the larger cells produce females, the smaller ones males. No pattern in the sequence or position of cells of the two sizes was detected.

Provisioning the Nest. Prey-laden females approach the nest only 10–15 cm high. When followed by a satellite fly, they frequently fly slowly beyond the nest using an undulating flight 2–6 cm high, the fly following slightly below and behind the wasp. Often they fly off a considerable distance and return, sometimes several times, before finally reaching the nest without the fly.

At the entrance the closure is removed with a few thrusts of the front legs, and the wasp enters quickly. Bees are brought in at fairly infrequent intervals; for example, one female was watched from 1405 to 1505 hours and brought in only three bees. Only 20 seconds to 3 minutes are spent inside the nest between trips for prey. Bees are stored partway down the burrow, 9–17 cm deep as measured from the surface directly above, well beyond the beginning of the steepened portion of the burrow. Prey consisted exclusively of small bees at all study sites, and with one exception all were Halictidae (Table 7-4).

Table 7-4. Prey records for Philanthus lepidus

Species	Central New York[1]		Eastern Massachusetts[2]	
ANDRENIDAE				
Pterosarus andrenoides	1 ♀			
HALICTIDAE				
Augochlora pura	1 ♀			2 ♂
Augochlorella striata	12 ♀	8 ♂	5 ♀	5 ♂
Augochloropsis metallica	1 ♀			1 ♂
Dialictus apertus				1 ♂
Dialictus coeruleus	2 ♀	1 ♂		1 ♂
Dialictus cressonii	15 ♀		7 ♀	6 ♂
Dialictus delectatus		20 ♂		1 ♂
Dialictus disabanci	4 ♀		1 ♀	
Dialictus heterognathus	6 ♀	2 ♂		
Dialictus imitatus	7 ♀		16 ♀	1 ♂
Dialictus laevissimus		10 ♂		3 ♂
Dialictus lineatulus	2 ♀			
Dialictus nigroviridis	1 ♀			
Dialictus oblongus	2 ♀			
Dialictus rohweri	4 ♀		1 ♀	
Dialictus tegularis			7 ♀	
Dialictus versans			8 ♀	
Dialictus spp.	5 ♀	20 ♂	3 ♀	6 ♂
Evylaeus divergenoides				3 ♂
Evylaeus macoupinensis		3 ♂		2 ♂
Halictus ligatus			1 ♀	
Lasioglossum coriaceum		1 ♂		
TOTALS 23 species (bees)	63 ♀	65 ♂	49 ♀	32 ♂

1. Granby Center and Rensselaerville
2. Bedford and Lexington

Natural Enemies

Satellite flies (*Senotainia trilineata* and *S. vigilans*) were abundant at all study sites and were often seen trailing prey-laden females. However, the devious flight patterns described above were apparently often successful, as only 3 of 40 cells excavated at Rensselaerville appeared to have had the contents destroyed by maggots (no such damage was found at other sites).

In Lexington, Massachusetts, the mutillid wasp *Dasymutilla nigripes* was common, and females of this species were sometimes seen entering accessory burrows and digging in them briefly. Since *D. nigripes* is a known parasite of *P. gibbosus* (Shappirio, 1948) it seems probable that it sometimes develops at the expense of *lepidus*. At Rensselaerville, *Dasymutilla gibbosa* was common in the nesting area, but we have no firm evidence that it was attacking *Philanthus*.

Philanthus bilunatus Cresson

P. bilunatus ranges throughout the eastern United States and southern Canada west to the Rocky Mountains. It is structurally similar to lepidus but appears less characteristic of open, sandy soil and more likely to occur in firm, coarse, sandy loam in partially wooded areas. It is evidently univoltine throughout its range and, like lepidus, is characteristic of late summer, from late July through early September. Evans and C. Lin (1959) reported on four nests in Ithaca, New York. We report here on four nests found near Togue Pond, Baxter State Park, Piscataquis County, Maine, on 7 August 1972, and on several nests and male territories found 23 km west of Livermore, Larimer County, Colorado, 23 July through 2 September 1984–1985. In the Rocky Mountains the species occurs at elevations from 2000 to 2600 m. Both sexes are common visitors to the blossoms of goldenrod (*Solidago* spp.). Krombein (1936) recorded the species from *Achillea*, *Chrysanthemum*, and *Daucus carota*.

Male Behavior

We have found only two male territories, both on the west side of a small juniper tree in a montane valley near Livermore, Colorado. We do not know the source of these individuals, as no nests could be found within 0.5 km of the site. On 11 August 1984 there were two territories, 1.5 m apart. One was on a small side branch of the tree 0.5–1.0 m above the ground. The male at this site was seen pursuing another male and butting it in the air at 1205 hours. The resident was watched for 10 minutes, 1211–1221, during which time it scent-marked 24 times, always on the tips of juniper twigs, going first up and then down over a distance of 2–4 cm. It spent considerable time flying slowly about among the branches, and each scent marking was preceded by such a flight. All perches were on juniper twigs. During the 10-minute period, a bee fly was pursued a short distance, but there were no other interactions with insects. This territory was not found to be occupied on subsequent days.

On the other hand, the second territory was occupied on 11, 12, 14, and 17 August (by unmarked males). This territory was close to the ground on a small currant bush (*Ribes cereum*) growing beneath the branches of the juniper. The resident male perched on currant leaves, on juniper twigs, and on nearby grasses, and all of them were scent-marked, as were a dead stick and a nearby sage. On grass blades, the wasp went up and down, covering 3–8 cm at each marking, but on currant leaves it went back and forth along the edge of the leaf. In a 10-minute period (1125–1135) on 12 August, there were 31 markings, some in bouts of 2; there were 10 interactions with intruding males, including one grapple to the ground. Most interactions consisted of pursuit, the resident following the intruder a few centimeters behind and below.

Similar observations were made at this territory on 14 and 17 August. The rate of scent marking on these dates was 42 and 32 per 10-minute period. In 40 minutes of observation over 4 days on two territories, we obtained a mean scent-marking rate of 3.2 per minute. Although males were seen on flowers up to 29 August, no further territorial behavior was observed. It is possible that this area was a resource for females, as it evidently bore honeydew; numerous bees and wasps flew about the tree, including both male and female *P. bilunatus*. On similar dates in 1985, the tree was not attractive to Hymenoptera, and no territories were seen here.

Nesting Behavior

Evans and C. Lin (1959) found one nest in flat soil and three in a vertical sandbank in central New York. The four nests found in Maine were in a hard-packed sandy road in an opening in spruce forest. The nests found near Livermore, Colorado, were in sloping soil along a road cut some 0.5 km from the territories described above. The soil here was a sandy loam on the surface but consisted largely of firm, degraded rock below 5–10 cm. Two nests were located here in 1984, 10 in 1985. *P. pulcher* nested in this same bank earlier in the season, but there was no seasonal overlap between the two species.

All nests observed in Maine and Colorado had small, unleveled mounds measuring 5–10 cm long and 3–7 cm wide. Of the four nests in Maine, two had one accessory burrow and two had two; these burrows measured from 0.5 to 1.5 cm in length and were dug close to the entrance of the true burrow and at a 40–90° angle to it. All 10 nests studied in Colorado in 1985 also had accessory burrows at least part of the time, usually throughout their active period (the two studied in 1984 were not discovered until they were well advanced; they had no accessory burrows). Length of the burrows at this locality varied from 0.3 to 8.0 cm (mean = 2.6, N = 49). These figures include daily measurements of accessory burrows at each nest, which, as the following examples show, were sometimes variable.

Nest 2 had an accessory burrow 1.5 cm deep on the right side when discovered on 23 July. On 30 July this burrow measured 1 cm; but on 2 August it had been deepened to 2 cm, on 3 August to 4 cm, and on 8 August to 6 cm.

Nest 3 had an accessory burrow on the left 1 cm deep on 30 July. On 1 August this burrow had been deepened to 2.5 cm, and by 8 August to 4.5 cm; but on 18 August it was only 1.3 cm deep, and there was a second accessory burrow on the right 0.3 cm deep.

Nest 10 was the only one that lacked an accessory burrow initially (2–4 August); but on 8 August a burrow was dug 4 cm deep to the right, and it was maintained until 18 August.

It was common to see females entering accessory burrows and digging in them briefly, particularly following a nest closure. The presence of two accessory burrows was unusual (only 3 of 49 records from the two localities). True nest entrances were closed except when the female had entered with prey.

Nests are proclinate and progressive (Fig. 7-1B). The oblique burrow is about 4 mm in diameter. Cells are spaced 3–6 cm apart at distances of 11–27 cm from the entrance. The maximum number of cells found per nest was 3 in New York, 2 in Maine, and 6 in Colorado. The 3 cells found in Maine were 14–15 cm deep, while 31 cells from Colorado were 9–17 cm deep (mean = 12.8 cm). Cells measure about 11 by 16 mm. From 9 to 14 bees are provided per cell (mean = 11.3, $N = 8$). Prey consists primarily of small Halictidae (Table 7-5).

Natural Enemies

At Togue Pond, Maine, several satellite flies (*Senotainia* sp.) were observed. Females followed by these flies undertook devious flight patterns and often flew away and returned a few moments later with their prey, usually without the flies. This behavior was confirmed at the Colorado site, where it was noted that females often landed some distance away from the nest and remained motionless for several seconds before flying in a circuitous pattern only 5–10 cm high either to the nest or to a second such "freeze-stop." One female made four freeze-stops—one of 25 seconds 2 m from the nest, a second of 10 seconds 0.5 m from the nest, a third of 20 seconds 3 m from the nest, and a fourth of 5 seconds 0.3 m from the nest—before walking to the entrance. No satellite flies were seen on that occasion, and it appeared that similar behavior occurred with or without the presence of these flies.

Table 7-5. Prey records for Philanthus bilunatus

Species	Tompkins Co. New York[1]	Piscataquis Co. Maine	Larimer Co. Colorado
COLLETIDAE			
Colletes sp.			1 ♂
Hylaeus modestus	1		
ANDRENIDAE			
Pseudopanurgus sp.			3 ♀ 2 ♂
HALICTIDAE			
Augochlorella striata	2		
Dialictus cressonii			2 ♀ 9 ♂
Dialictus imitatus	5		
Dialictus ruidosensis			2 ♀ 4 ♂
Dialictus versans		1 ♂	
Dialictus zephyrus	2		
Dialictus spp.		5 ♀ 4 ♂	15 ♀ 11 ♂
Evylaeus spp.		1 ♀ 3 ♂	2 ♀ 16 ♂
Halictus confusus	6		1 ♀
Halictus ligatus	2		
Halictus sp.			1 ♀
Lasioglossum leucozonium	1		
Lasioglossum sisymbrii			1 ♀ 4 ♂
Sphecodes sp.		1 ♂	7 ♀ 4 ♂
TOTALS 17 species	19	6 ♀ 9 ♂	34 ♀ 51 ♂

1. Data from Evans and Lin (1959). No records were kept on the sex of prey.

In spite of this behavior, several cells (5 of 31) at the Colorado site were found to contain maggots or puparia; six other cells contained only bee remains and may also have had the contents destroyed by maggots. An adult *Senotainia trilineata* emerged on 1 May 1985 from a nest provisioned on 2 September 1984. Several larger, hole-searching flies, *Phrosinella* sp., were also observed at this site, and on 6 May 1986 two flies of this genus emerged from puparia taken from nests 23 July and 8 August 1985.

At this same site, a female *Dasymutilla vesta* was seen digging a series of shallow holes in the nesting area not far from known nests of both *bilunatus* and *pulcher*.

Philanthus solivagus Say

P. solivagus is restricted to the northeastern United States and southeastern Canada. It is a species of moderate size, on average slightly smaller than *sanbornii* and larger than *gibbosus*. Ristich (1956) studied the species briefly in Maine and reported on the success of miltogrammine parasites. Evans and C. Lin (1959) described nests and prey from near Ithaca, New York. Our discussion here includes additional data gathered near Rensselaerville, New York, in August and early September, 1970, as well as a few casual observations made in eastern Massachusetts.

P. solivagus is a wasp characteristic of late summer, appearing about the first of August and nesting well into September (the latest recorded nesting occurred on 21 September, at Bedford, Massachusetts). Most nests we have found were in man-made excavations in fairly firm, fine-grained sand, with sparse vegetation. While the species appears to prefer steep slopes or vertical banks, several nests near Ithaca and near Rensselaerville were in flat or nearly flat soil. Common associates are *gibbosus* and *lepidus*, although the former often nests in coarser soil and we have never found the latter in steep slopes. Both sexes of *solivagus* are frequent visitors to flowers, especially those of *Solidago*. Krombein (1936) also recorded the species from *Achillea millefolium*. We have made no observations on male behavior.

Nesting Behavior

While digging, females are rarely seen outside the entrance; they tend merely to scrape the sand behind them from the burrow so that it forms a mound just outside the entrance or rolls down the slope. Mounds at entrances of newly constructed nests may be quite large, as much as 7 by 10 cm and 2 cm deep in the center; unlike the mounds made by many other *Philanthus*, they are not grooved in front of the entrance. After a few days the mound tends to erode away, and the entrance may become rather large, as much as 15–20 mm in diameter. No accessory burrows have been observed in this species.

Nests are proclinate and progressive, with a serial arrangement of cells (Fig. 7-3). The burrow descends obliquely, forming an angle of 20–40° with the horizontal. At a distance of 15–50 cm from the entrance (usually about 35 cm) the burrow passes upward for 3–5 cm, then obliquely downward again for 5–10 cm before passing upward again, then once again obliquely downward. There may be three to eight such "serrations" before the burrow reaches the level of the first cells. The first cells are normally located 40–60 cm from the entrance, subsequent cells progressively deeper, some as much as a meter from the entrance. Each cell lies at the terminus of a short cell burrow, 5–15 cm long; cell burrows are irregularly spaced along the main burrow. Distance be-

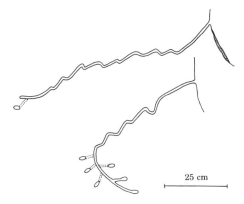

Figure 7-3. Two nests of *Philanthus solivagus*, showing the serrate profile of the burrow (Ithaca, N.Y.). Burrows shown with dashed lines were packed with soil and could not be traced accurately. (From Evans and C. Lin, 1959, courtesy of *Wasmann Journal of Biology*.)

tween cells varies from 3 to 10 cm. The cell burrows are closed off with a firm barrier of sand as soon as the egg has been laid.

The maximum number of cells found in any nest was 14 (5 September 1970, near Rensselaerville). In this nest and one other from the same area having 12 cells, the burrow appeared to have a second major branch beginning 15–40 cm from the entrance. It seems probable that each female maintains a single nest for her life span, but we have not followed marked females throughout the active season.

Evans and C. Lin (1959) reported cell size to be 12 by 20 mm, the number of prey items per cell 6–14. Studies near Rensselaerville demonstrated that in fact cells were of two sizes: small cells measuring about 10 by 16 mm and containing 6–8 prey items and larger cells measuring about 12 by 20 mm and containing 10–14 prey items. It is assumed that small cells produce males, large cells females. Cells of the two sizes appear to be made in a random pattern and sequence. Near Ithaca, the cells excavated were from 45 to 85 cm from the entrance (mean = 66 cm, $N = 17$). Near Rensselaerville, the cells excavated lay from 32 to 100 cm from the entrance (mean = 57 cm, $N = 31$).

Provisioning the Nest. The nest entrance is invariably left open while the female is outside. During the night and during periods when she is inside digging, the entrance is usually closed with a plug of sand. Provisioning females plunge into the entrance quickly from a height of 1–2 m. Prey consists of a wide variety of small bees as well as a few wasps (Table 7-6). In addition to the prey recorded in the table, Ristich (1956) has reported the following from Maine: Vespidae, *Eumenes* sp.; Sphecidae, *Ectemnius* sp.; Andrenidae, *Andrena* sp.; Halictidae, *Halictus*, 3 spp. Contents of a single cell are often diverse, one cell having

Table 7-6. Prey records for Philanthus solivagus

Species	New York: Ithaca[1]	New York: Rensselaerville	
VESPIDAE			
Ancistrocerus adiabatus	1	6 ♀	
Ancistrocerus catskill	2		
SPHECIDAE			
Astata nubecula	1		
Ectemnius continuus	1		
Ectemnius lapidarius	1		
Ectemnius maculosus		4 ♀	
Lestica confluenta	1		
COLLETIDAE			
Colletes americanus	1		
Colletes kincaidii		1 ♀	
Colletes simulans armatus		1 ♀	3 ♂
Colletes solidaginis			1 ♂
ANDRENIDAE			
Andrena asteris	1		
Andrena nubecula	1	1 ♀	
Andrena simplex	1	4 ♀	
Andrena subaustralis	1		
Andrena sp.	2		
HALICTIDAE			
Agapostemon texanus			1 ♂
Agapostemon virescens	11		
Augochlora pura	3		
Augochlorella striata	20	14 ♀	1 ♂
Augochloropsis metallica	15	3 ♀	7 ♂
Dialictus coeruleus			1 ♂
Dialictus cressonii		1 ♀	
Dialictus delectatus			1 ♂
Dialictus laevissimus	2		
Dialictus lineatulus	1		
Dialictus oblongus	1		
Dialictus rohweri		1 ♀	
Dialictus versans	1		
Dialictus zephyrus		1 ♀	
Dialictus spp.		2 ♀	
Evylaeus cinctipes		7 ♀	1 ♂
Evylaeus truncatus		1 ♀	
Halictus ligatus	37	16 ♀	4 ♂
Halictus confusus	5		
Halictus rubicundus	1	2 ♀	5 ♂
Lasioglossum coriaceum	31	1 ♀	14 ♂
Lasioglossum leucozonium	1		
Sphecodes davisii	1		
ANTHOPHORIDAE			
Epeolus scutellaris			1 ♂
TOTALS 40 species (7 wasps, 33 bees)	144	66 ♀	40 ♂

1. Data from Evans and Lin (1959). No records were kept on the sex of the prey from this locality.

been found to contain eight species of bees. The time required to capture bees varies considerably. One wasp brought in only two bees in 1 hour, while another brought in five prey items in 30 minutes (Evans and C. Lin, 1959).

Females have been observed bringing in prey as late as 1800 hours. They usually remain within the nest for 3–5 minutes, then peer out the entrance for a moment before taking flight. Near Ithaca, we found only one bee that appeared to have been stored in the burrow, and it was our impression that bees were usually taken directly to a cell and that more than one cell might be stocked before eggs were laid. Excavations near Rensselaerville did not confirm this impression. In one nest we found 11 bees apparently stored in the burrow at a vertical depth of 48 cm, 37 cm from the entrance. There was an empty cell some distance beyond this, to which the bees would presumably have been carried. The other 13 cells in this nest contained eggs, larvae, or cocoons, except for 1 cell that had been parasitized by miltogrammine flies.

Natural Enemies

Prey-laden females are very frequently followed by one to three satellite flies, *Senotainia trilineata*. Ristich (1956) reported two cases in Maine in which flies of this species had deposited maggots on the prey; in both cases two maggots had been deposited, in one instance on the middle femur, in the other on the dorsal surface of the abdomen. *S. trilineata* females were commonly seen trailing wasps near Ithaca, and it was observed that the wasps often flew about deviously in an effort to shake off the flies. Similar observations were made near Rensselaerville; in this locality only 3 of 33 cells excavated appeared to have been parasitized. In one instance the maggots were reared successfully but proved to be Phoridae of the genus *Megaselia*. The affected cell was excavated on 5 September and found to contain about 20 small maggots; adult flies emerged 18–25 September. Thus it does not appear that *Senotainia* is a particularly successful parasite of *solivagus* despite its abundance in the field. Ristich also noted that these satellite flies are less successful in attacking *solivagus* than in attacking *Aphilanthops frigidus*. This lack of success is perhaps a consequence of the fact that nest entrances are not closed, permitting the wasp to enter very quickly.

Near Ithaca, we also noted the miltogrammine *Phrosinella fulvicornis* at a nest entrance, but we have no data on the incidence of parasitism by that species.

Philanthus albopilosus Cresson

P. albopilosus is a widely distributed species that occurs from the Atlantic coast west to Alberta, Utah, and Arizona and from southern Canada to northern Mexico. It is, however, very local in distribution, being confined to sand dunes and blowouts. There are three published reports on its behavior, a very brief one by Evans and C. Lin (1959) followed by longer reports by Evans (1975) and by Hilchie (1982). Hilchie described a separate subspecies for the more extensively maculated southern populations, calling it *P. albopilosus manuelito*, but the color differences appear no greater than those occurring in many other *Philanthus* that have justifiably not been split into subspecies.

P. albopilosus is restricted to broad expanses of fine-grained sand, chiefly on windward slopes and in blowouts, where the sand is reasonably firm. *P. psyche* often occurs in these same areas but is commonly found on the periphery, where there is sparse vegetation. The nests of *albopilosus* tend to occur in more central areas, where there is little or no vegetation and where high midday surface temperatures and a certain amount of blowing sand are common features. Only a few other insects occupy this habitat, notably the sand wasps *Bembix pruinosa* and species of *Microbembex*.

Evans (1975) reported on aggregations near the eastern and western extremities of the range. Near Albany, New York, a population was studied 7 July–13 August 1970 in a slightly sloping man-made excavation into dunes. Near Roggen, Weld County, Colorado, a population was studied in blowouts 8 July–10 September 1974, with a few additional observations on similar dates over the following few years. Several additional observations were made near Hasty, Bent County, Colorado, near Socorro, New Mexico, and near Monahans, Ward County, Texas, all in areas of sand dunes. Evans and C. Lin (1959) reported on a nest excavated near Tuba City, Arizona. Hilchie (1982) studied a population in a dune field near Empress, Alberta, during the summer of 1977.

In Texas and New Mexico, the species is active as early as 12 June, but we have not encountered it before early July in Colorado or New York. There are evidently two generations in the latter two states, a second generation appearing in late August and early September after a midsummer hiatus. There may well be more than two generations farther south. At sites close to the Mexican border, for example, adults have been collected as early as May and as late as October (Hilchie, 1982).

Male Behavior

Males were abundant at the sites in New York and near Roggen, Colorado. They typically perch on the sand and less frequently on a pebble or stick, assuming an "alert" posture, with antennae extended rigidly forward, head slightly elevated, and abdomen curved slightly upward posteriorly. The front of the head is densely clothed with silvery pubescence, which glitters in the sun and may represent an important signal in the spacing of the males. Distance between perching males is rarely less than 1.5 m except very temporarily. When males perch more closely than this, and when a male flies close to another male, the result is a brief encounter during which the wasps butt one another repeatedly while rising in a spiral to a height of 0.5–1.0 m. Males often fly toward other passing insects without actually butting them and sometimes fly up and land elsewhere in the absence of any obvious stimulus.

Although we did not mark any males or perches, we obtained the impression both in New York and in Colorado that the wasps moved readily from one point to another and that in fact there were no favorite perches occupied by the same or different males over a period of time. Rather, males maintained a certain distance between themselves by a more or less random series of moves from one part of the sand surface to another. Use of a marking pheromone, as described for several species of *Philanthus*, seems most unlikely in this species. Indeed, the clypeal and abdominal brushes so characteristic of males of this genus are absent in *albopilosus*, and the mandibular glands are small, about the size of those of females (Fig. 2-4).

On several occasions we saw males descend on females working at their nest entrances or fly after females carrying prey, but in no case did mating ensue. We did observe one mating pair, at 1015 hours on 17 July 1975, near Roggen, Colorado. The two were motionless on the sand in the nesting area for at least 30 seconds, facing in opposite directions. We did not see them initiate copulation and thus have no data on its duration. Hilchie (1982) also observed copulating wasps "resting on the dune surface" (p. 99).

Nesting Behavior

Nests are widely spaced, usually 1.0–1.5 m apart and in no case closer than 0.8 m. Nests dug in slopes face into the slope, whereas those on flat or nearly flat sand face in various directions. On one occasion, near Roggen, we found 26 active nests in early September in an area measuring 3 by 30 m. On another occasion, in July, we found about 50 nests in an area about 8 by 20 m in a large blowout.

Females typically begin a nest in the morning (0900–1100) and dig intermittently throughout the day. When they stop digging and fly off, as they do periodically, they invariably close the entrance of the hole. Digging wasps appear just outside the entrance from time to time and scrape the sand into a mound. Since they do not emerge more than about 1 cm from the entrance, but tend to turn from side to side slightly while clearing, the resulting mound is close to the entrance and somewhat circular in outline. Mounds in New York measured from 11 to 16 cm wide (mean = 13.2) and from 9 to 12 cm long (mean = 10.8, $N = 4$). In Colorado, mounds measured 12–15 cm wide (mean = 13.5) and 15–28 cm long (mean = 19.0, $N = 7$). In both areas the maximum depth of mounds approximated 1 cm. The appreciably greater length of mounds in Colorado doubtless reflected the much greater depth of the nests there (see below).

Following completion of the nest, the female backs somewhat farther from the entrance (3–6 cm) and, still facing it, uses a series of zigzag movements to level off the portion of the mound close to the entrance. This behavior requires several minutes and results in the mound's assuming a roughly crescentic shape. The flattened area close to the entrance typically contains a somewhat star-shaped pattern of lines, formed when the female makes the initial and later closures by scraping sand over the entrance from all sides. The initial closure is especially prolonged; in making it, the female often excavates soil from the ends of the radiating lines, resulting in the formation of one or more accessory burrows. Of 5 newly completed nests studied in New York, 2 had one accessory burrow each, both burrows measuring about 1.5 cm deep. Of 34 newly completed nests studied near Roggen, Colorado, 27 had one or more accessory burrows, which varied from 0.5 to 7 cm deep (the majority were 0.5–2.0 cm deep). Of the 27 nests with such burrows, 14 had one, 6 had two, 3 had three, 2 had four, and 2 had six (Figs. 7-4, 7-7). These burrows were constructed immediately after completion of the nest, at the time of initial closure, and were not renewed later. Nests more than a few hours old rarely showed any evidence of accessory burrows, as they tended to be filled by wind action. Hilchie (1982) observed accessory burrows in Alberta but did not provide data on their depth or incidence. In no cases were wasps seen closing these burrows, although a closure of the true burrow was maintained at all times.

Figure 7-4. Mound at entrance of a nest of *Philanthus albopilosus*, Weld Co., Colo. Point of the pencil overlies the closed nest entrance; the openings of five accessory burrows are visible.

Nest Structure. The nest of *albopilosus* is of unique structure, although basically declinate (Figs. 7-5, 7-6). It consists of a nearly horizontal section (which we term the *vestibule*) that runs close beneath the surface and leads to a fairly vertical section (the *gallery*) that penetrates the soil deeply. The vestibule passes only 2–5 cm beneath the surface and is commonly of greater diameter than the gallery, often 5–12 mm compared with 4 mm for the gallery. Whether this greater diameter results from cave-ins from the ceiling, which are removed by the wasp, or from actual excavation is not clear. Presumably nests are dug only where the soil has a slight crust sufficient to maintain a ceiling for a burrow so close to the surface. Although in flat soil the vestibule dips slightly (up to an angle of 10° with the surface), when the burrow is dug into a sloping bank the vestibule may actually slant upward slightly. The vestibule is of appreciable length, about the same in New York and western localities; in contrast, the cells averaged about 2.5 times as deep in western localities, with no overlap in the range of variation (Table 7-7).

Although we have spoken of the gallery as being nearly vertical, in some nests it curves down gradually before becoming essentially vertical. Cells are constructed at the ends of rather long (5–20 cm) cell

Other North American Species of *Philanthus* **187**

Figure 7-5. Initial, oblique burrow (vestibule) of a nest of *Philanthus albopilosus*, Weld Co., Colo. Ruler is about 16 cm long.

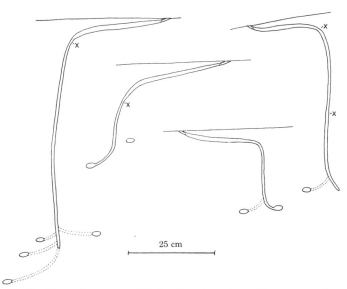

Figure 7-6. Profiles of four nests of *Philanthus albopilosus*. The nests at the extreme left and right are from near Roggen, Weld Co., Colo.; the two shallower nests in the center are from near Albany, N.Y. Cell burrows, shown by dashed lines, had been filled and could not be traced accurately. The position of prey stored in the burrow is indicated by an *x*. (From Evans, 1975, reprinted with permission from *Annals of the Entomological Society of America*, © 1975, Entomological Society of America.)

Table 7-7. Nest dimensions (cm) of Philanthus albopilosus (mean and range of variation)

Locality	No. nests	Length of vestibule	Length of gallery	Depth of cells
New York: near Albany	7	24.6 (16-31)	18.7 (11-22)	19.2 (13-25)
Colorado: near Roggen	9	24.5 (18-35)	47.8 (23-70)	52.0 (31-62)
Colorado: near Hasty	4	24.5 (16-40)	50	52
New Mexico: near Socorro	1	20	40	48
Texas	1	40	49	--

burrows that branch from near the end of the gallery. Since these cell burrows are closed as soon as the cell is provisioned, they can rarely be traced. Indeed, the cells of any one nest tend to be spaced so widely that it is difficult to locate all of them. The maximum found in any one nest was four, but it seems likely that more are sometimes made. Cells measure 12–15 mm long by 7–8 mm high.

Evidently females of this species prepare relatively few cells per nest but make a series of nests during their lifetimes. At the site in New York, we marked five nests in the process of being dug on 8 August. On 11 August, two of them were no longer active, and on 13 August all appeared to be inactive. Of 26 nests marked near Roggen, Colorado, on 6 September, only a very few were still being provisioned on 10 September. Hilchie (1982) reported that the nests he studied were active for only one or two days.

It should be noted that *Bembix pruinosa*, which commonly nests close beside *albopilosus* in many dune areas, also makes a long preliminary burrow close beneath the surface before penetrating the soil deeply (Evans, 1957b). However, this burrow is filled in when the gallery is dug, and a new entrance is made at some distance from the original entrance. That is not the case in *albopilosus*, which continues to use the original entrance. It might be noted, too, that *B. pruinosa* also tends to make slightly deeper nests in Colorado than in New York (cell depth averaging 35 and 27 cm, respectively). The contrast is not nearly so great as in the case of *P. albopilosus*, however.

Provisioning the Nest. Females prey on a variety of small bees and occasionally on small wasps (Tables 7-8, 7-9). They approach the nest swiftly, flying only 3–6 cm above the sand, holding the paralyzed prey

Table 7-8. Percentages of wasps and bees taken as prey by Philanthus albopilosus

Locality	% Wasp individuals	% Wasp species	% Bee individuals	% Bee species
New York: near Albany	17	36	83	64
Colorado and New Mexico[1]	1	11	99	89
Alberta: near Empress[2]	38	59	62	41

1. Weld and Bent Counties, Colo., and Socorro County, N. Mex.
2. Data from Hilchie (1982).

with their middle legs. At the entrance, the closure is removed with several strokes of the forelegs and the wasp enters, sliding the prey backward slightly and holding it with the hind legs, in the usual manner of philanthine wasps (Fig. 7-7). Miltogrammine flies were not abundant at any study site, and we saw no evidence of the evasive flights characteristic of many *Philanthus* species.

Provisioning occurs at a slow pace, one female (New York) bringing in two bees over a 45-minute period. Hilchie (1982) found rates of provisioning to be highly variable, one female returning in 5 minutes with prey, another in 55 minutes. Prey is not taken directly to a cell but is left near the far end of the vestibule, near the point where the gallery angles down sharply (in two nests we found bees in the gallery itself, in one case at considerable depth). Approximately enough bees are stored here to provision one cell; when enough have accumulated, the female stops provisioning and spends a long time inside the nest, removing the prey to a cell much deeper in the soil.

Within the cell, prey are piled mostly venter-up, with the egg laid longitudinally on the topmost prey, as usual in this genus. The number of prey items per cell varied from 7 to 10 in New York (mean = 8.3, $N = 6$) and from 6 to 8 in Colorado (mean = 6.7, $N = 6$), doubtless a reflection of the somewhat smaller prey employed at the New York site, such as *Dialictus* and *Oxybelus*. Similar data are not available from other sites.

Hilchie (1982) observed a female capturing prey in Alberta. As he described it, "A beewolf cruising about 50 to 60 cm above the sand near a colony of *Microbembex monodonta* (Say) wasps, dived to the ground, attacked and captured one of the *Microbembex*. The latter individual was either flying near or resting on the sand" (p. 104).

Hilchie found that most prey items varied in length from 8 to 11 mm

Table 7-9. Prey records for Philanthus albopilosus

Species	New York: near Albany	Colorado & New Mexico[1]	Alberta: near Empress[2]
ICHNEUMONIDAE			
Genus and species?			1 ♀
VESPIDAE			
Ancistrocerus catskill	1 ♂		
Stenodynerus anormis			1 ♂
SPHECIDAE			
Aphilanthops frigidus	1 ♂		17 ♂
Crabro denningi			1 ♂
Diodontus sp.			1 ♀ 3 ♂
Mellinus abdominalis			1 ♂
Microbembex monodonta			1 ♀ 1 ♂
Mimumesa fuscipes	1 ♀		
Oxybelus bipunctatus	1 ♀		
Oxybelus uniglumis	5 ♂		
Podalonia violaceipennis			1 ♂
Tachysphex tarsatus			1 ♀
Tachysphex exsectus			1 ♂
Tachytes sp.		1 ♂	
COLLETIDAE			
Colletes hyalinus		21 ♂	
Colletes simulans armatus	1 ♂		
Colletes spp.		1 ♀ 5 ♂	19 ♂
ANDRENIDAE			
Andrena robertsonii	1 ♀		
HALICTIDAE			
Agapostemon angelicus		31 ♂	
Agapostemon sp.			1 ♂
Dialictus imitatus	1 ♀		
Dialictus lineatulus	1 ♂		
Dialictus pilosus	2 ♀ 6 ♂		
Halictus confusus	2 ♀ 2 ♂		
Halictus ligatus	1 ♀	21 ♂	
Halictus rubicundus	17 ♂		1 ♀ 23 ♂
Lasioglossum leucozonium	9 ♂		
Lasioglossum sisymbrii		3 ♂	
Lasioglossum sp.			2 ♂
Sphecodes sp.			3 ♂
MELITTIDAE			
Hesperapis sp.		1 ♀	
MEGACHILIDAE			
Osmia sp.			1 ♀
TOTALS 33 species (15 wasps, 18 bees)	9 ♀ 43 ♂	2 ♀ 82 ♂	6 ♀ 74 ♂

1. Weld and Bent Counties, Colo., and Socorro County, N. Mex.
2. Data from Hilchie (1982).

Figure 7-7. A female *Philanthus albopilosus* entering her nest with a bee. Openings of two accessory burrows can be seen in the foreground.

and that 92.5% were males. The percentage of males was also very high at the aggregation in New York (83%) and at those in Colorado and New Mexico (98%). Wasps provided a considerably higher proportion of the prey in northern localities than in Colorado and New Mexico, doubtless a reflection of the fact that in the areas studied there were fewer bees in the immediate vicinity of the dunes (Table 7-8). Hilchie (1982) provided a long list of potential prey, that is, bees and wasps of suitable size taken in sweeps around the dunes.

Natural Enemies

Near Albany, New York, a female was seen to enter a hole about 15 cm from her nest, apparently by accident. This hole was occupied by a tiger beetle larva (*Cicindela* sp.), which promptly seized the wasp and began to feed on it. At this same locality, 1 of 11 cells excavated contained maggots that had consumed most of the prey. These maggots formed their puparia a few days later and the following spring gave rise to adult flies, *Phrosinella fulvicornis* (Miltogramminae). No natural enemies were observed at the study sites in western states, and Hilchie (1982) noted none in Alberta.

Chapter 8
A Brief Review of Eurasian Species

Although there are more than a hundred species of *Philanthus* in Eurasia and Africa, most of them remain unstudied. We include here, for comparative purposes, a brief survey of knowledge of five species that have been studied to varying degrees.

Philanthus triangulum (Fabricius)

There is an extensive literature on the widely distributed, moderately large European and African wasp *P. triangulum*, which has achieved notoriety as a predator of honey bees (Chapter 2). Some of the earlier reports are in part erroneous, for example, Fabre's (1891) statement that provisioning is progressive. Fortunately some excellent detailed studies of the behavior of this species have been published within the past few decades. We make no attempt to review all of the earlier publications, but the accounts by Hamm and Richards (1930), Olberg (1953), and Grandi (1961) are especially noteworthy and contain bibliographies; Olberg's account is replete with excellent photographs. Tinbergen and his associates have done much important work on prey capture and orientation (Tinbergen, 1932, 1935; Tinbergen and Kruyt, 1938; Beusekom, 1948), and Rathmayer (1962a, 1962b, 1966) has made a careful study of prey paralysis. Recent studies by Simonthomas and his coworkers have clarified many details of both male and female behavior (Simonthomas, 1966; Simonthomas and Simonthomas, 1972;

Simonthomas and Poorter, 1972; Simonthomas and Veenendaal, 1978). Some of this research has already been alluded to in our general survey of the biology of *Philanthus* (Chapter 2), and we here review only what appear to be some of the more distinctive features of *triangulum*. An early report on the African subspecies *triangulum diadema* (Mally, 1909) suggests that there are no important differences between it and the European subspecies.

P. triangulum inhabits sandy soil, such as partially vegetated sand dunes, sandy woodlands, and artificial fills or excavations. In the northern half of Europe, populations fluctuate greatly in size from year to year, perhaps in response to winter temperatures or to the number of sunny days in the summer. Fabre's studies and many of those of Simonthomas were made in southern France; marked yearly fluctuations may also occur here, as a result of parasite pressure (Simonthomas and Simonthomas, 1972). In large aggregations, nests tend to be spaced at least 10 cm apart; when closer than that, the females tend to fight, seizing one another with their mandibles, though without causing injury (Simonthomas and Simonthomas, 1972).

Male Behavior

Males usually spend the night in short burrows, but early in the season they may share nesting burrows with other members of both sexes (Bouvier, 1916). It is probable that these are temporary associations of siblings that have emerged from the parental nest. Life of a male centers around three places: the sleeping burrow, the territorial site, and the feeding site (in Simonthomas's studies, chiefly *Thymus serpyllum*, *Rubus ulmifolius*, and *Frangula alnus*). These three places may be some distance apart (10–90 m). All sites tend to be near but somewhat apart from the female nesting site, but occasionally sleeping burrows or territories have been found close to female nests; in one case a nest was found inside a territory (Simonthomas and Poorter, 1972).

Near Naboude, France, male territorial perches were found to be situated on pine trees from 0.5 to 4.0 m tall and also occasionally on heather or bracken. Simonthomas and Poorter marked numerous males and found them to be quite site-specific; these researchers present a table showing the occupancy of more than 30 territories over a period of 3 weeks. In one case a territory was occupied by the same male for 11 of 13 successive days, but in other cases one male replaced another. Visiting males, often more than one at a time, were present at many territorial stations, resulting in frequent skirmishes. Territorial males spend much time perching, their antennae raised slightly. From time to time a male "moves hurriedly up and down grass blades, pine needles, or branches, meanwhile flickering its antennae and tipping the substrate with them" (Simonthomas and Poorter, 1972, p. 149). Dr. Simonthomas kindly loaned us a motion picture he had made of this behavior, and it compares very closely to the scent-marking behavior of several North American species. This correspondence is confirmed by the studies of Borg-Karlson and Tengö (1980) and the photographs in their paper. Simonthomas and Poorter (1972) observed three copulations, all at territorial stations.

Nesting Behavior

From 1 to 3 days are required for completion of a nest. Sometimes a female usurps a nest, driving away the occupant, who rarely succeeds in recovering the nest. Joint occupancy of a burrow by provisioning females has not been reported, although Simonthomas and Simonthomas (1972) saw one female take a bee into a nest that was not her own. According to these authors, the female usually occupies the same nest for her entire lifetime (usually 5 or 6 weeks). However, Bouvier (1900) reported that females make three successive nests, and Tinbergen (1932) reported that females leave the nest after having made four or five cells and dig a new nest, usually close by.

The soil at the nest entrance is not leveled, the mound serving as a proximate cue in nest finding. The mound tends to be somewhat fan-shaped, as the female backs out when removing soil from the nest, not always following the same path. We know of no records of the construction of accessory burrows by *triangulum* females. Various authors have reported that the nest entrance is closed by the female when she leaves the nest, the closure varying from thorough to hasty and incomplete. In their studies in southern France, however, Simonthomas and Simonthomas (1972) found that females rarely closed their nests. During one season, seven females never closed their nests, five did so only 3% of the time, and one closed 40% of the time.

The burrow enters the ground obliquely, then after 15–25 cm levels off. The horizontal portion is lengthened from day to day as cells are added and eventually reaches considerable length. Total length of the burrow in its final form sometimes exceeds a meter. Simonthomas and Simonthomas report that a second horizontal part is sometimes constructed from the end of the oblique burrow. There are two generations a year in southern France, and these authors suggest that members of the second generation may reoccupy first-generation nests, a condition paralleling that in *gibbosus* in North America (Evans, 1973). Nests are proclinate and progressive, cells being constructed in a rather regular pattern from short cell burrows off the main burrow, successively farther from the entrance. Up to 14 cells have been reported per nest.

Provisioning the Nest. Hunting occurs principally on flowers. Tinbergen (1932) spent much time watching a hive of honey bees but failed to observe prey capture there. On the other hand, at the Dakhla Oasis in Egypt, females commonly hunt at beehives (Simonthomas and Simonthomas, 1977). This behavior occurs chiefly in the winter, when few flowers are in bloom in the area. The beewolves alight near the hive and capture returning foragers. They are no match for guard bees at the entrance, which may attack and even kill the wasps if they have an opportunity. Nevertheless, beewolves often capture enough foragers to seriously diminish the amount of food reaching the colony. Hunting at hives has also often been reported in Europe.

Although worker honey bees constitute the usual prey of this species, other bees of similar size are occasionally taken. They include species of *Halictus*, *Andrena*, and *Megachile*. Mimetic *Eristalis* flies are not accepted as prey (Fahringer, 1922). The number of bees per cell varies from one to six. Simonthomas and Simonthomas (1972) found that 41% of more than 500 cells they excavated contained two bees; 19% contained only one bee; 22%, three; 10%, four; 7%, five; and 1%, six.

Natural Enemies

The cuckoo wasp *Hedychrum intermedium* is a major parasite of *triangulum* (Olberg, 1953; Simonthomas and Simonthomas, 1972). These wasps may be very abundant, averaging as many as 2.6 near each host nest in France. They appear to be attracted to both closed and open nests and may oviposit on the prey as it is brought in or after it has been placed in the nest. The *Philanthus* female may attack the parasite, but she is unable to sting it because of its thick integument and its behavior of rolling into a ball. Simonthomas and Simonthomas report the decline or near extermination of several nesting aggregations as a result of the attacks of cuckoo wasps.

Quite a number of other insects are reported to attack *triangulum*, including other species of cuckoo wasps as well as several flies, including the miltogrammine *Metopia leucocephala* and conopids of the genus *Physocephala* (Hamm and Richards, 1930). Simonthomas and Simonthomas (1972) found that nearly 10% of bees brought in were later dragged out of the nest; about a third of these proved to have been parasitized by flies. Olberg (1953) provided several excellent photographs of various natural enemies. The mutillid wasp *Dasylabris maura* is also recorded as a parasitoid of *triangulum* (Simonthomas and Simonthomas, 1980).

Philanthus coronatus Fabricius

P. coronatus ranges throughout warmer parts of the Palaearctic region. It is a large species, females averaging about 17 mm in length. Tsuneki (1943) studied *coronatus* in a chestnut plantation near the Seoul Airport in Korea. He found the nests to be rather shallow, the oblique burrow descending to a depth of only 16 cm, with a total burrow length of only 37 cm. Three cells were located at the ends of short cell burrows that branched from the main burrow. Cells measured 3 cm long by 2 cm in diameter and were provisioned with six to eight bees each. The mound of earth at the nest entrance was not leveled, and there was a short accessory burrow on each side of the nest entrance; these burrows were separated from the entrance by 5–8 cm and were 4–5 cm long. The accessory burrows were not closed, but the main burrow was closed whenever the female left the nest. Prey consisted of female halictid bees. Berland (1925) had earlier reported Halictidae as prey of this species.

Philanthus venustus Rossi

P. venustus also occurs in the southern parts of the Palaearctic region, but it is a smaller species than *coronatus*, females averaging about 12 mm in length. It was first studied by Ferton (1905) in southern France, where he found it making deep nests in sandy soil. Prey consisted of small bees, identified chiefly as Halictidae, which were stored in various parts of the burrow before being taken to a cell. Ferton found several barriers of sand in the burrow separating caches of paralyzed bees.

Grandi (1961), who studied *venustus* in Italy, found the burrows to be rather shallow. Burrow length was only 15 cm, and the terminus 7–8 cm deep. The mound at the entrance was not leveled and an outer closure was maintained. Grandi found males and females in some abundance on flowers of umbelliferous plants. He found four species of Halictidae to be used as prey. Prey listed by Grandi and by Ferton as belonging to the genus *Halictus* are all small species and would currently be placed in other genera, such as *Lasioglossum* and *Evylaeus*. Krombein (1972) reported *Evylaeus interruptus opacus* as prey in Greece. He found individuals of both sexes on a sandy beach at Tolon, Argolis.

Philanthus basalis basalis Smith

Krombein (1981) has studied the Asiatic species *P. basalis basalis* briefly in Sri Lanka. It is a specialist on workers of the Indian honey bee, *Apis cerana indica*. Nests are dug into nearly vertical banks of soft sandstone; they occur in small aggregations with entrances at least 15 cm apart. Burrows are sometimes occupied by more than one female, although Krombein evidently did not observe more than one female provisioning the same nest. The burrows penetrated the bank more or less horizontally and often had sharp upward or lateral bends. Krombein believed that old cells were cleaned out and reused and that provisioning was progressive, but these would be unusual features for members of this genus, and his observations are not sufficiently detailed to be convincing on these points. Mites belonging to a genus near *Vidia* (Winterschmidtiidae) were apparent scavengers in the wasps' cells.

Philanthus pulcherrimus Smith

The Indian species *P. pulcherrimus* was the subject of a brief report by Dutt (1914), who found females nesting in long tunnels in sandy banks. Bees were captured on flowers and belonged to the genera *Halictus*, *Ceratina*, and *Apis*.

Chapter 9

An Overview of Male Mating Strategies

Many field ethologists have been attracted to the conspicuous and fascinating behavior patterns of female digger wasps, but for a long time they paid scant attention to male behavior. In fact, William Morton Wheeler (1919), an otherwise astute observer of insect behavior, referred to the male of Hymenoptera as "an ethological nonentity" (p. 3). Recent studies reveal this as a premature judgment and have stimulated interest in the behavior of male digger wasps. We know of no papers published before N. Lin's (1963) study of *Sphecius speciosus* that concentrated on the behavior of males. In contrast, we are aware of 26 published since then that deal solely with males. Male *Philanthus* figure in more than half of these reports (14) and have been the subject of some of the most detailed studies. This research has been complemented by the growth of a similar body of data on males of other families of aculeate bees and wasps (see Thornhill and Alcock, 1983).

Recent interest in male behavior has also been influenced by the revival of sexual selection theory since the early 1970s. Trivers (1972) elaborated on Darwin's (1859) original concept of sexual selection, noting that females generally invest more time and energy in each offspring than do males and thus have a relatively limited fecundity. Males of most species are free from parental duties and substantial nutrient investment in gametes, so their reproductive success is essentially determined by the number of eggs they fertilize. Therefore, females (i.e., their eggs) become the resource that limits male fitness. Competition for this resource may result in a high variance in reproductive success among males in a population. Darwin realized that this competition could manifest itself in two ways. (1) Males will attempt to gain *possession* of receptive females or of locations likely to be frequented by them. (2) Males will attempt to gain the *attention* of females by advertising qualities that are likely to make them relatively better mates. The components of the process of natural selection resulting from these two forms of competition are now referred to as *intrasexual* and *intersexual selection*, respectively. This simple argument provides the logical basis for a myriad of testable predictions concerning the evolution of male mating systems (Thornhill and Alcock, 1983).

This chapter summarizes our knowledge of the behavior of male beewolves, makes comparisons with other species of digger wasps, and proffers some general conclusions concerning the evolution and adaptive nature of mating strategies in *Philanthus*. We have chosen to include a broad discussion of male digger wasps because, owing to a paucity of information, previous major comparative works on wasps contained no such discussions (e.g., Evans, 1957b, 1966c; Iwata, 1976). Throughout the discussion, we refer to the species listed in Tables 9-1 and 9-2; references not cited in the text can be found in those tables. Although this review is not exhaustive, it contains all of the major studies of male behavior that we know of, as well as many papers that include observations on males as part of more general reports. When possible, we evaluate the available information in the context of sexual selection theory, as detailed in Alcock et al., (1978), Emlen and Oring (1977), and Thornhill and Alcock (1983). Because of the nature of the available evidence most of the discussion refers primarily to adaptations apparently evolved through the process of intrasexual selection. Where we have given free rein to speculation, we hope it is obvious that we have done so. We also hope that others will be tempted to follow up on our suggestions.

Male Participation in Nesting Activities

As a general rule, although female digger wasps invest heavily in offspring, the activities of males are aimed primarily at obtaining copulations; they contribute nothing but genes to their offspring. When not engaged in mate-seeking activities they are feeding on nectar or resting. The exceptions—the cases in which male behavior, either directly or indirectly, helps the female to rear offspring—seem restricted to very few genera, with the best-known examples in *Oxybelus* and *Trypargi-*

lum (Hook and Matthews, 1980). In some species of these genera, the male remains in or near the nest entrance of a provisioning female and mates repeatedly with her. The female lays an egg on the last prey placed in each cell, so the male that engages in such behavior increases the probability that his sperm will fertilize that egg, assuming last-male sperm precedence. While at a nest entrance, the male chases away all intruding insects, including conspecific males and parasitic flies. Consequently, in *O. subulatus* (Peckham, 1977), nests guarded by males have nearly half the rate of miltogrammine cleptoparasitism as guarded nests. In such cases, the protection of the nest may be best considered a coincidental result of activity evolved to exclude conspecific male competitors from the immediate vicinity. In *Trypargilum*, however, some aspects of male behavior appear to be true examples of parental effort. As in *Oxybelus*, the male remains with a single nest and mates repeatedly with the provisioning female. The male not only repels potential parasites but assists in other activities related to nest maintenance and provisioning. In at least one species, *T. superbum*, the male continues to guard the nest after the female has completed it and moved on, leaving only after all the larvae have spun cocoons (Coville and Griswold, 1984). The importance of male guarding behavior to the females is evident in the observation that females of *T. tenoctitlan* do not provision a nest until they have attracted mates (Coville and Coville, 1980). The examples from *Trypargilum* and *Oxybelus* remain the only ones in which male activities have a demonstrated influence on components of female reproductive success, although observations of males of *Tachytes distinctus* (Lin and Michener, 1972), *Dynatus nigripes*, and some *Lindenius* species suggest similar effects (Hook and Matthews, 1980). One might speculate that the territorial activities of male *P. bicinctus* and *P. psyche* around nests reduce the incidence of nest parasitism. Considering the fact that nearby nests may be provisioned by females with whom the male has not mated, however, male behavior has undoubtedly evolved strictly in the context of mating effort (in the sense of Low, 1978).

The behavior of males of species such as *T. superbum* may increase confidence of paternity, but at the expense of decreasing potential mating success, because these males spend considerable time with a single female. In contrast, male beewolves, when successful at obtaining a mate, spend as little as 5–10 minutes with her before resuming the search for another female. Thus, although differences do arise within the Sphecidae, Trivers's (1972) basic argument concerning the disparity in parental investment between the sexes and the consequent higher *potential* reproductive success of individual males certainly applies to most species of digger wasps.

Patterns of Female Receptivity

The females of some species of digger wasps (e.g., *Bembecinus quinquespinosus*, *Bembix rostrata*, *Cerceris watlingensis*, *Oxybelus sericeus*, *Sphecius speciosus*, and *Trigonopsis cameronii*) are known to be receptive immediately upon emergence from their nest cells at the beginning of the season. In some studies (e.g., of *P. bicinctus*, *P. crabroniformis*, and *P. zebratus*), although the exact time at which the female emerged was unknown, dissections of recently mated females revealed that their ovaries had not fully developed, indicating recent emergence. For *P. bicinctus* and *P. zebratus*, it was found that marked females mated close to the time at which they initiated their first nests.

Although further studies are required, circumstantial evidence suggests that females of most digger wasps mate only once in their lives (Alcock et al., 1978). In *Philanthus*, most copulations are observed in the first 2–3 weeks of the season, although nesting may continue for at least several weeks after that (e.g., in *bicinctus*). Most contacts between males and females do not lead to copulation (e.g., Gwynne, 1980; O'Neill, 1979). Males and females rarely interact on flowers, where they commonly feed together. Provisioning females usually vigorously refuse mating attempts, an easy task, as they are usually larger than the males. Rebuffs of male suitors are observed in other digger wasps as well. Thus, after apparently mating only a single time (or at most several times) early in their nesting sequence, female digger wasps do not require a further supply of sperm. Their relatively low fecundity (O'Neill, 1985) and the fact that males probably do not pass nutrients to them during insemination mean that females may gain little advantage from multiple mating. The obvious exceptions to the generalization that female digger wasps mate just once are *Oxybelus* and *Trypargilum*. Hook and Mathews (1980) suggested that the advantages that the females of these two genera gain from the presence of the male at the nest has been an important factor in selecting for female acquiescence to repeated copulation attempts. Alcock et al. (1978) discussed the potential costs and benefits associated with multiple mating in aculeates.

Seasonal Patterns of Activity

Males' tendency to emerge before females and their shorter seasonal activity period suggest that their activity patterns are synchronized with the presence of newly emerged or nesting females, which apparently mate only once. Protandry, emergence of males before females, appears to be a general phenomenon in digger wasps (Tables 9-1 and 9-2). In most hole-nesting species protandry in each nest is necessary because male cells are closest to the entrance (Krombein, 1967). In *Philanthus* and other ground nesters, to which this restriction does not apply, protandry is usually not pronounced. Rather, there is a relatively long seasonal overlap in emergence of the sexes. Furthermore, even though male beewolves begin to emerge before females, they do not set up territories until the females have begun to nest (e.g., Gwynne, 1980; O'Neill, 1981). Both sexes emerge over a 2–3-week period, as indicated by emergence trap data (Gwynne, 1980) and the fact that fresh-looking territorial males and nesting females appear over a prolonged period.

The reasons for prolonged emergence and protandry are unclear. Staggered emergence may result from the influence of variation in temperatures within and between cells on rates of development. Protandry may be related to the shorter pupal period of male aculeates (Krombein, 1967), which are usually smaller than conspecific females (O'Neill, 1985). We have also found protandry in the digger wasp *Bembecinus quinquespinosus*, a species lacking sexual size dimorphism, however. Protandry is predicted in species whose females mate only once and are receptive at emergence or shortly thereafter (Wiklund and Fagerström, 1977). Given the prolonged emergence period of females and the relatively short life span of males, however, the lack of pronounced protandry in many digger wasps is not surprising. Males emerging too early might not survive until the peak of female emergence. A model of the timing and form of the male emergence schedules developed by Iwasa et al. (1983) suggests that males that emerge after the first females have emerged but before the peak of female emergence can realize a fitness equal to that of early emergers, even though they may miss the opportunity to mate with early-emerging females.

The seasonal activity period of male digger wasps is generally shorter than that of females. Examples of species in which males disappear or obviously decline while females are very active and abundant include many *Philanthus*, most *Bembix* (Evans, 1957b), *Ammophila aberti* (Hager and Kurczewski, 1985), and *Tachysphex* (Kurczewski, 1966). In

Bembecinus quinquespinosus, males disappear nearly completely before most females have even begun to nest (Evans et al., 1986). Freeman (1980) has shown that males of *Sceliphron assimile* have a lower mean survivorship than do females. We suspect that a similar difference applies to *Philanthus*.

Protandry and prolonged emergence schedules can affect the level of competition among males. A good estimate of the level of competition when the emergence sex ratio is approximately unity is the *operational sex ratio*, or OSR (Sutherland, 1985), the ratio of mate-seeking males to receptive females at a given time. As the OSR increases, the intensity of sexual selection increases (Emlen, 1976). Thus, if we can determine the OSR, we can estimate the level of competition among males. In beewolves, the staggered emergence schedule of females means that only a small proportion will be receptive on any given day. This fact, combined with the fact that the peak of male emergence precedes the peak of female emergence, should result in a high OSR during parts of the season. No studies of beewolves have set out to make precise determinations of the OSR, which could be roughly estimated as the ratio of the number of males that have emerged *through* a given date to the number of females emerging *on* that date. Data for *bicinctus*, which has an emergence sex ratio of approximately 1:1 (Figure 3 in Gwynne, 1980), indicate that the OSR is relatively high throughout the period over which receptive females are present, a 2-week period near the middle of the nesting season. More accurate estimates of the OSR, which include such information as the survivorship schedule of males, would be valuable in future studies.

Spatial Distribution of Receptive Females and Mate-Searching Males

A variety of sites at which mating typically occurs have been identified for various digger wasps: foraging areas (including nectar sources and hunting sites), sleeping aggregations, emergence sites, single nests and aggregations of nests, and flyways between areas visited by females (Alcock et al., 1978; Thornhill and Alcock, 1983).

With few exceptions, foraging areas containing nectar sources or prey appear to be either widely scattered or abundant; these areas do not represent good locations for territories, because they are unlikely to contain concentrations of receptive females. This is true for *Philanthus*; females of most species prey primarily on solitary bees or on a variety of bees and wasps (O'Neill and Evans, 1982), which, being generalists themselves, are found on a variety of flowers (Eickwort and Ginsberg, 1980). As mentioned, although male and female *Philanthus* commonly feed together on flowers, mating attempts have not been observed there. Mating with females at foraging sites is a common strategy of male bees (Eickwort and Ginsberg, 1980), but only a few cases have been documented among the Sphecidae, and none of them involve territoriality (Table 9-1).

Although nocturnal sleeping aggregations are common in digger wasps (e.g., Linsley, 1962), mating does not usually occur there. The one documented exception involves *Steniolia obliqua*.

In some species emergence sites are localized because of colonial nesting in the previous generation. Thus, males can find emerging females by searching for them in the area in which they themselves emerged. Furthermore, cues may be available that allow them to home in on the approximate or even the exact point of emergence of individual females. Males of *Sphecius speciosus* establish territories near conspicuous emergence holes. Males of *Trigonopsis cameronii*, *Bembecinus tridens*, *B. quinquespinosus*, and *Bembix rostrata* detect females before the latter emerge from the ground; odor has been implicated as the important cue in the latter species.

For most digger wasps, including *Philanthus*, it does not seem useful to distinguish between emergence sites and nest sites, since there is broad temporal and spatial overlap between emerging and nesting females (i.e., females often nest in the same area from year to year). Furthermore, females may still be unmated when they nest shortly after emergence. Thus, because the most favorable time to locate a receptive female seems to be near the beginning of her nesting activities, it is often to a male's advantage to center his mate seeking in or around nest sites. Mating at or near nesting areas is well documented within the Sphecidae.

For a male beewolf, dense nesting aggregations are a mixed blessing. They provide locations where territories can be established in the vicinity of numerous receptive females, but they also attract other males, increasing the level of competition for limited perch sites. In *Philanthus*, variation in nest density among species and even populations is considerable (Table 9-2). This fact presents male *Philanthus* with wide variation in potential for locating and monopolizing areas likely to contain receptive females. We expect a male to be territorial in a given area only if the net gain from such behavior is greater than the net gain from an alternative tactic (Brown, 1964; Alcock et al., 1978).

The following list describes four general categories of nest density in terms of approximate number of active nests per potential male territory (see Tables 9-1 and 9-2). Keep in mind that these categories, in reality, form part of a continuum of possibilities.

(1) *Extremely clumped nest distribution.* Five or more active nests would likely be within a territory of typical size (approximately 1–2 m in diameter at most). Among beewolves, *pulcher* and *zebratus* fall within this category (Figures 3–9, 5–1). In the Jackson Hole population of *pulcher*, only a few territories would have fit in the compact nest sites. In the nearby population of *zebratus*, not more than 8–10 territories would have fit in among the nests. We estimate that many fewer than 10% of the males in each population could have been simultaneously territorial there. The resulting competition for the limited number of sites would have been extremely high, and each male would have expended much time and energy in defense of the perch. In fact, males of both these populations set up territories near but not within the nest sites, avoiding some of the cost of defense yet remaining near receptive females. Other beewolves, including *crabroniformis* and *triangulum*, may also fall within this category in terms of both nest density and territory location. We know of no examples of digger wasps that typically set up territories within such dense nesting areas.

(2) *Clumped nest distribution.* Two or fewer nests would be in a territory of typical size, with many other nests within 5–10 m of the perch; females would commonly pass near the territory while going to and from nests. As nest density is reduced somewhat from that described in Category 1, there is less competition for the increased number of perch sites. In the beewolves *bicinctus* and *psyche*, males establish territories among nests of this density. The relationship between nest and territory location is best documented in *bicinctus* (Gwynne, 1980). Territory location varied with nest distribution both between and within years, and there was a significant correlation between the number of days a territory was occupied by males and the number of nests present within 3 m.

Many nonterritorial digger wasps have nest densities that fit Categories 1 and 2 (Table 9-1). In most such species studied, males patrol nonaggressively through the nesting/emergence areas in search of receptive females. Two species of beewolves have such strategies: *albopilosus* and *zebratus*. In *albopilosus* at the Roggen, Colorado, site, males patrolled among female nests located in the open sand of a blowout. Large males of *zebratus* in the Jackson Hole population patrolled for females in the airspace above the nesting area, intercepting

206 The Natural History and Behavior of North American Beewolves

Table 9-1. Species of digger wasps in which males patrol for receptive females

	Nest density[1]	Protandry (no. days)	Aggression among males?
1. Males patrol emergence area			
Bembecinus quinquespinosus	1	yes (2)	yes
Bembecinus tridens	1	yes	yes
Bembix rostrata	1-2	yes (2)	yes
Microbembex monodonta	1	yes	-
2. Males patrol nesting/emergence area			
Alysson melleus	2	-	-
Ammophila aberti	1-2	yes (3)	yes
Ammophila azteca	2	-	-
Ammophila harti	2	yes (3-15)	yes
Bembecinus nanus	2	-	no
Bembix nubilipennis	1	-	yes
Bembix multipicta	1-2	-	-
Bembix pruinosa	1-2	-	no
Bicyrtes quadrifasciata	3	-	-
Cerceris truncata	1	-	-
Cerceris watlingensis	-	-	no
Entomognathus memorialis	-	-	yes
Lindenius columbianus	2	-	yes
Miscophus californicus	1	-	-
Oxybelus bipunctatus	1	yes	yes
Philanthus albopilosus	2	-	yes
Philanthus zebratus (some)	1	yes (1-2)	no
Plenoculus boregensis	2	-	-
Podalonia valida	4	-	no
Stictia carolina	1-2	yes	no
Tachytes intermedius	1	yes (2-8)	-
3. Males patrol foraging area			
Cerceris graphica	-	-	no
Ammophila procera	1	-	-
4. Males patrol at sleeping clusters			
Steniolia obliqua	1	-	no

1. Numbers in this column refer to categories discussed in the text under "Spatial Distribution."

females making orientation flights or returning to nests from foraging areas.

(3) *Nests somewhat localized.* No nests would be found within a territory of typical size, but some would commonly lie within 10-20 m. Traffic of females through the territory would be infrequent. This category fits species such as *barbiger* and *serrulatae*, whose females tend to nest in the same local area from year to year, but at low densities. Males of these two beewolf species establish territories in the general vicinity

Larger males more successful?	Mating observed?	Alternative tactics reported?	References
yes	yes	yes	O'Neill & Evans 1983b
-	yes	no	Lüps 1973
-	yes	yes	Schöne and Tengö 1981
-	yes	no	Evans 1966c
-	yes	no	Evans 1966c
-	yes	no	Powell 1964
-	no	no	Evans 1965
-	yes	no	Hager & Kurczewski 1985
-	yes	no	Evans & O'Neill 1986
-	yes	no	Evans 1966c
-	no	no	Evans 1966c
-	yes	no	Evans 1957b
-	yes	no	Evans 1966c
-	no	yes	Werner 1960
-	yes	no	Elliott 1984
-	yes	no	Miller & Kurczewski 1972
yes	no	no	Miller & Kurczewski 1973
-	no	no	Powell 1967
-	yes	yes	Peckham et al. 1973
-	yes	no	Evans 1975
-	yes	yes	O'Neill & Evans 1983a
-	yes	no	Rubink & O'Neill 1980
-	yes	no	Steiner 1975; Evans & O'Neill, unpublished
-	yes	no	Evans 1966c
-	yes	no	Kurczewski & Kurczewski 1984
-	no	no	Alcock & Gamboa 1975
-	yes	no	Bohart & Knowlton 1953
-	yes	no	Evans 1966c

of nests but without the smaller-scale correlations between nest and territory location evident in *bicinctus* and *psyche*.

(4) *Nests dispersed*. Nests are fairly evenly dispersed within the habitat frequented by the species, so that territories may be some distance from the nearest nest. Lack of aggregations makes female nest and emergence locations unpredictable for males. We have found this type of nest distribution difficult to document, although we believe that it is common in many species of *Philanthus* (e.g., *barbatus* and *biluna-*

Table 9-2. Species of digger wasps in which male territoriality has been reported

	Nest density[1]	Protandry (no. days)	Territories aggregated? (no.)	Larger males more successful?	Territories scent-marked?	Mating observed?	Alternative tactic reported?	References[3]
1. Defense of single nests								
a. Mate with emerging females								
Trigonopsis cameronii	3	–	–	–	–	yes	no	Eberhard 1974
b. Mate with nesting female								
Dynatus nigripes	–	–	–	–	no	yes[2]	no	Kimsey 1978
Oxybelus sericeus	–	–	–	–	no	yes[2]	yes	Hook & Matthews 1980
Oxybelus subulatus	–	–	–	–	no	yes[2]	no	Peckham et al. 1973
Trypargilum politum	–	–	–	–	no	yes	no	Hartman 1944, Cross et al. 1975
Trypargilum rubrocinctum	–	–	–	–	no	yes	no	Paetzel 1973
Trypargilum spinosum	–	–	–	–	no	yes[2]	yes	Hook 1984
Trypargilum tenoctitlan	–	no	–	yes	no	yes[2]	yes	Coville & Coville 1980
2. Defense of emergence area								
Sphecius speciosus	1	yes	yes (1-8)	–	no	yes	no	N. Lin 1963
Sphecius grandis	–	yes	yes	yes	no	yes	no	Alcock 1975e, Hastings 1986
Stictia vivida	–	–	yes	–	no	no	no	Evans 1966c
3. Defense associated with nesting/emergence area								
a. Nests commonly within territory								
Philanthus bicinctus	2	yes (6-7)	yes (2-43)	yes	yes	yes	yes	Gwynne 1978, 1980
Philanthus psyche	2	–	yes (>30)	yes	yes	yes	no	O'Neill 1979, 1983a
Tachysphex terminatus	1-2	yes	yes (23-50)	–	no	yes	no	Kurczewski 1966
b. Dense nest aggregation nearby								
Philanthus crabroniformis	1-2	–	yes (2-10)	yes	yes	yes	no	Alcock 1974, O'Neill 1983a, Evans & O'Neill

Species								References
Philanthus pulcher	1	yes (1-2)	yes	yes	yes	yes	yes	O'Neill 1983a, Evans & O'Neill
Philanthus triangulum	1	yes (2-10)	yes	-	yes	yes	no	Simonthomas & Poorter 1972; Borg-Karlson & Tengö 1980

c. Five or more known nests nearby

Philanthus zebratus (some)	1	yes (1-2)	yes	yes	yes	no	yes	O'Neill & Evans 1983a
Philanthus barbatus	-	-	yes (1-2)	-	yes	no	no	Evans 1982
Philanthus barbiger	3	-	yes (1-3)	-	yes	no	no	Evans & O'Neill
Philanthus multimaculatus	-	-	yes (>8)	-	yes	yes	no	Alcock 1975d
Philanthus serrulatae	3	-	no	-	yes	no	no	Evans & O'Neill

d. Nests widely scattered

Philanthus basilaris	4	-	yes (4-47)	yes	yes	yes	no	O'Neill 1983b

4. Territorial, but nest distribution uncertain

Cerceris nigrescens	-	-	no	-	yes	no	no	Evans & O'Neill 1985
Clypeadon taurulus	-	-	yes (8-12)	-	yes	yes	no	Alcock 1975c
Eucerceris arenaria	-	-	yes (2-15)	-	yes	no	no	Alcock 1975c
Eucerceris canaliculata	-	-	yes (3-8)	-	yes	no	no	Alcock 1975c
Eucerceris cressoni	-	-	no	-	yes	no	no	Evans & O'Neill 1985
Eucerceris flavocincta	-	-	yes (2-30)	yes	yes	no	no	Steiner 1978, Evans & O'Neill 1985
Eucerceris rubripes	-	-	yes	-	yes	no	no	Alcock 1975c
Eucerceris superba	-	-	no	-	yes	no	no	Evans & O'Neill 1985
Eucerceris tricolor	-	-	yes	-	yes	no	no	Alcock 1975c
Philanthus bilunatus	3	-	yes (1-2)	-	yes	no	no	Evans & O'Neill
Philanthus gibbosus	3	-	yes (2-4)	-	yes	no	no	Alcock 1975a
Philanthus pacificus	-	-	-	-	yes	no	no	Evans & O'Neill
Philanthus tarsatus	-	-	yes (1-5)	-	yes	no	no	Evans & O'Neill
Philanthus ventilabris	-	-	yes (2)	-	yes	no	no	Alcock 1978
Tachytes distinctus	-	-	-	-	no	no	yes	Lin & Michener 1972

1. Numbers in this column refer to categories discussed in text under "Spatial Distribution."
2. In these species (only), multiple mating by individual females has been observed.
3. References to Evans & O'Neill (without date) refer to information presented in this book.

tus), *Cerceris*, and *Eucerceris* (e.g., *flavocincta*). We believe, too, that we can confidently place *basilaris* in this category (O'Neill, 1983b). At Great Sand Dunes, males of this species establish territories in groups equivalent in many ways to the *leks* of many vertebrates. The territorial areas do not contain unusual quantities of resources or nest substrate.

In species with nest densities that fit into Categories 3 and 4 (e.g., *Bicyrtes quadrifasciatus* and *Podalonia valida*), males often patrol broad areas of habitat in search of females, because female location is unpredictable and because in many of these species males apparently cannot attract females to scent-marked territories (see below). Unfortunately, their dispersed spatial distribution makes these species difficult to study.

Considerable variation in nest density has been noted in some species of *Philanthus*, especially *zebratus*, *barbatus*, and *pulcher*. In *zebratus*, we noted no patrolling flights in a population whose nests were somewhat dispersed, although such flights occurred in a much denser aggregation not many kilometers distant. In *barbatus*, one site had a group of aggregated nests with territories close beside it, while at another site (again not far away) nests were very widely spaced and territories aggregated far from known nests. Territories within a sparse nesting aggregation were noted in *pulcher* at one site, although most aggregations of this species are dense (Category 1) and territories are peripheral to the nesting area.

Nest density is, of course, a rough correlate of the real variables of interest, the number of receptive females appearing per unit area per unit time and their locatability (Alcock et al., 1978). Therefore, it is not surprising that, although nest density is a general correlate of territory location, it is a weak predictor of the specific mating tactic evolved by any given species. Although extremely high nest density probably precludes territoriality in the nesting area as a male mating tactic, for example, several other tactics are available. In some species with such nest distributions, males patrol for females at the nest site (e.g., many *Bembix*); in others, males set up territories just outside the nest site (e.g., *crabroniformis* and *pulcher*); and in one (*zebratus*), males in the same population do both.

Among species of beewolves and other philanthines, the density of territories, like the density of nests, varies considerably. In *barbatus*, *barbiger*, *bilunatus*, and *serrulatae*, males either set up territories isolated from others or form small groups of several territories. Thus, interterritory distance is often quite great. In other species—*basilaris*, *bicinctus*, *crabroniformis*, *psyche*, and *pulcher*—territories are typically aggregated. In *bicinctus* and *psyche*, this grouping of territories is a result of attraction of males to nest sites. In *basilaris, crabroniformis*, and *Eucerceris flavocincta*, such aggregations do not seem to be a fortuitous result of the presence of females, localized resources, or topographical features critical for attracting females. Rather, the males appear to be attracted to each other for some reason, and females visit the area because males are present. Perhaps when territories are grouped the chemical signal that attracts females (see below) is enhanced in a manner beneficial to individual males. Female choice may be responsible for the evolution of grouped territories, if females only visit aggregations of males in order to compare potential mates (Bradbury, 1981; O'Neill, 1983b).

Behavior of Territorial Males

Scent-Marking Behavior and its Function

Chapter 2 discussed scent-marking behavior, its morphological correlates, and the chemistry of mandibular gland secretions. As noted, the hypothesized function of the volatiles is to act as a sex pheromone. The differences observed in pheromone chemistry between two North American species are interesting in this regard. The extracts studied were collected from sympatric populations where territories of the two species were intermingled with each other and with territories of a third species (*psyche*). Thus, some degree of reproductive isolation may be afforded by the specific pheromone blends. Other sites with sympatric multispecies groupings of territories are listed in Table 9-3.

We have talked about scent marking as if the sexual attractant function were well documented. In fact, a valuable piece of information is lacking; no bioassays have been done to verify that females are attracted by the chemicals. Some observational evidence supports the sex pheromone hypothesis, however. We have watched females of *basilaris*, *psyche*, and *pulcher* approaching territories just before mating with the resident males. In all cases they were flying directly upwind, often in an obviously zigzagging flight pattern. That is the behavior expected of an insect orienting to a windborne pheromone (Shorey, 1977).

A major alternative hypothesis is that the pheromones act as territory markers, warning conspecific males that a perch is occupied. This theory seems unlikely, given the large number of aggressive interactions on territories. Rather, males may be attracted to territories *be-*

Table 9-3. Groups of Philanthus species occurring in very similar habitats at several localities

Location	Species
Roggen, Colo.	albopilosus, psyche, tarsatus
Pawnee Grasslands, Colo.	basilaris, multimaculatus, serrulatae
Chimney Rock, Colo.	barbiger, basilaris, bicinctus
Deadman's Bar, Jackson Hole, Wyo.	pacificus, zebratus
Huckleberry Hot Springs, Wyo.	crabroniformis, pulcher, zebratus
San Rafael Desert, Utah	barbiger, multimaculatus, psyche, and others

cause of the chemicals deposited on the plants. Usurping a territory that has already been scent-marked may save them some of the energy cost of producing pheromone or allow them to be territorial if their own supplies of the chemicals are depleted (Alcock, 1975c; O'Neill, 1983a). Vinson et al. (1982) have shown that mandibular gland size in scent-marking territorial males of the bee *Centris adani* decreases dramatically across the daily period of territoriality.

Scent marking is an obligate component of territoriality in philanthines. Males begin to scent-mark within several minutes of arriving on their territories each day and continue to scent-mark until shortly before leaving. The rate of scent marking varies across species, however. In some (e.g., *pulcher*) the rate is fairly constant throughout the day, whereas in others (e.g., *basilaris*, *bicinctus*, and *multimaculatus*) males scent-mark at a higher rate soon after they have arrived each day. Species also differ in the overall rate of scent marking (Table 9-4). Much of the variation can be explained in terms of variation in body size among species. There is an inverse correlation between body size and the rate of scent marking, perhaps simply because larger species have larger clypeal hair brushes and so are able to deposit more pheromone each time they scent-mark. There seems no reason to believe that variation in the distance over which females must be attracted or variation in territory group size explain variation in the rate of scent marking. Males of *basilaris*, for example, scent-mark at a much lower rate than do males of *psyche*, a smaller species, even though *basilaris* territories are usually farther from nests than are *psyche* territories. Fur-

Table 9-4. Rates of scent-marking in 12 North American species of Philanthus[1]

Species	Size category	Number males observed	Total obs. time (min.)	Mean no. scent marks per min.
barbiger	1	5	72	3.0
politus	1	1	22	2.0
psyche	1	20	100	1.5
serrulatae	1	3	65	2.2
bilunatus	2	4	40	3.2
pulcher	2	5	426	1.1
tarsatus	2	2	151	1.7
barbatus	3	2	20	1.4
crabroniformis	3	7	66	2.3
basilaris	4	9	1337	0.2
zebratus	4	8	199	0.7
bicinctus	5	6	360	0.4

1. Spearmann's rank correlation for size rank versus rate of scent marking: $r_s = -0.69$, $P < 0.02$.

thermore, males of both species establish territories in groups of comparable size.

Several lines of evidence indicate that territorial males adjust their scent-marking activities to variation in both wind direction and wind speed. (1) Males of most species observed (*basilaris* is an exception) primarily scent-marked plants upwind of their perches (e.g., O'Neill, 1979, 1981). Furthermore, when wind direction changed, as it often did in areas frequented by *barbiger*, *psyche*, and *pulcher*, the males responded either by moving their perches so that they remained downwind of the marked plants or by scent-marking a different set of plants on the new upwind side of the territory. (2) Males of many species precede scent marking with a weaving flight, in which they fly up from the perch and rapidly zigzag back and forth directly in front of and downwind of the scent-marked stems (O'Neill, 1979; Borg-Karlson and Tengö, 1980). It has been hypothesized that this flight allows the male to monitor either pheromone levels within the territory or wind speed (O'Neill, 1979). (3) The mean wind velocity when males of *psyche* scent-marked was nearly twice the overall mean wind velocity during the same period (O'Neill, 1979). (4) Aggregations of territories of *basilaris* males at Great Sand Dunes assume an elliptical shape with the long axis approximately parallel to wind direction. Although such wind-oriented activities are expected if the males are scent-marking plants to produce wind-borne pheromone signals, they cannot be completely explained at this point. They may be related to monitoring of pheromone levels within territories, identification of optimal wind velocities at which to scent-mark (Sower et al., 1973), and positioning of territories downwind of those established by other males in order to take advantage of the other males' pheromone deposition (O'Neill, 1983b).

Males of all species of philanthines in which territoriality has been observed scent-mark (Table 9-2). The behavior has been reported in species of four genera (*Philanthus*, *Eucerceris*, *Cerceris*, and *Clypeadon*) of widespread placement within the subfamily (Alcock, 1975c, 1975d; Gwynne, 1978; Evans and O'Neill, 1985), so it is apparently a relatively ancient characteristic in this group. This conclusion is substantiated by the fact that males of only 2 of 11 genera within the subfamily Philanthinae lack clypeal hair brushes (Bohart and Menke, 1976). Scent-marking behavior and the form of the associated morphological structures are fairly consistent within the genus *Philanthus*, although abdomen dragging and hair brushes have secondarily disappeared in *albopilosus*. Differences do occur with other genera of philanthines. The form of abdomen dragging is different and often less conspicuous in other species (e.g., *Eucerceris flavocincta*, Evans and O'Neill, 1985, and *Clypeadon* spp., Alcock, 1975a; O'Neill, personal observation). The form of the hair brushes in *Philanthus* is also different from that in both *Eucerceris* and *Cerceris* (Bohart and Menke, 1976; Evans and O'Neill, 1985).

Although territoriality is widespread within the Sphecidae, neither abdomen dragging nor any other obvious form of scent marking has been observed in species of subfamilies other than the Philanthinae (Table 9-2). Certainly, it would not be expected in territorial species of *Oxybelus* and *Trypargilum*, in which males are closely associated with individual nests, or in species such as *Sphecius speciosus*, in which territories are set up where females have no choice but to enter them (i.e., the females emerge from the ground within territories). Scent marking to attract receptive females gives philanthine territoriality a great deal of flexibility and utility relative to that in other sphecids. Territories need not be restricted to areas where females will probably occur in great numbers, as in most territorial digger wasps (e.g., *Oxybelus*, *Sphecius*, *Stictia*, *Tachysphex*, and *Trypargilum*). This characteristic may be of particular value to males of species in which nest distribution is variable or unpredictable (e.g., *barbatus*, *basilaris*, *pulcher*, and *zebratus*) or in which some females do not nest in aggregations (e.g., *bicinctus* and *pulcher*).

Several problems arise when attempts are made to explain the evolution of scent marking in philanthines. First, the females of insects usually signal to males, thus avoiding the energy cost and risk of searching for males (Thornhill and Alcock, 1983). Females of *Philanthus* may be rendezvousing with signaling males while engaged in other activities (nesting, hunting, searching for flowers, etc.), however; so they may not be traveling far out of their way to mate (Alcock et al., 1978). Females of *bicinctus* and *psyche* fly to and mate at territories within several meters of their nests. In other species (e.g., *basilaris* and *pulcher*), the female may be attracted to males in the general hunting area or in flyways between foraging and nesting areas. Thus, female beewolves may actually benefit from not undertaking the loss of time and energy associated with signaling to males.

Second, how can we explain the need for signaling in species in which the males are already very close to the females (Gwynne, 1978)? Scattered nest distribution may well be the ancestral condition for female philanthines. Scent marking and long-distance attraction of females may have originally evolved in the context of spatially separated nests with locations unpredictable to males. Its continued presence in species that nest in aggregations remains a puzzle (Gwynne, 1978). Spatial and long-term temporal variation in nest density, as well as the unexplored role of female mate choice, may select for its continued presence. Any scenarios concerning the evolutionary history of scent marking remain necessarily speculative at this time. An intriguing possibility is that the male sex pheromone originated as a mimic of prey odors (J. O. Schmidt, personal communication, 1986).

Male Aggressive Interactions on Territories

Along with scent marking, aggressive interactions are the most conspicuous activities of territorial males. The level of aggression varies among species, from the noncontact interactions of male *psyche* at Roggen to escalated combat involving butting and grappling in most species.

The primary question to ask about a territorial species is this: *What is being defended at such a cost of time and energy and presumably at some risk of injury?* In some cases it is probably the location itself that is valuable and worth defending or usurping. Males of *bicinctus* and *psyche* defend small portions of a nesting area where a limited number of sites are available. Occupation of these territories places males close to receptive females. In *bicinctus*, the more frequently occupied perches

Figure 9-1. Territorial defense and scent marking by *Philanthus bicinctus*. (Drawing by Byron Alexander.)

are those near the densest aggregation of nests. Similarly, territories of male *Sphecius speciosus* tend to be close to the portion of the emergence area with the largest number of emergence holes. In other species (e.g., *barbiger* and *pulcher*), although territories may be associated with nesting sites, we can assign no absolute or relative value to specific perches. *P. pulcher* territories, for example, are established in small light-colored patches of soil near nesting areas. Although such locations are abundant, male *pulcher* fight vigorously to usurp or maintain possession of sites. The same holds for males of other species, including *basilaris*, in which territories are not associated with nest aggregations or important patches of resources.

Something is present on territories of all philanthines that should be of value to any male that can usurp: scent-marked plants. Assuming that the pheromone has some nontrivial cost of production, that it can be produced at a limited rate, or that storage of large amounts of the chemicals presents a problem, there is an advantage in defending a scent-marked territory (Alcock, 1975c; O'Neill, 1983a). Observations of the temporal dynamics of territorial areas each day support this hypothesis. Early in the day, many males attempt to usurp scent-marked territories, even though many perches occupied on previous days are empty. This behavior has been observed in species that differ from one another in the spatial relationships between nests and territories (e.g., *basilaris*, *psyche*, and *pulcher*). Furthermore, males of some species (e.g., *basilaris*, *bicinctus*, and *multimaculatus*) are known to scent-mark at a higher rate earlier in the day. A male usurping a territory soon after this disproportionately high investment has been made can remain territorial for the rest of the day at a lower cost.

In areas where the territorial areas of two or more species overlap, we have not seen prolonged aggressive interactions between males of different species. In fact, on 19 occasions during a study at Great Sand Dunes in 1980, males of *bicinctus* occupied the same territories used at other times by males of *basilaris*; however, only intraspecific aggressive interactions were observed on these territories. We take this as further evidence that the presence of the pheromone may be more important than location in determining the value of the territory to conspecifics.

The Significance of Intraspecific Variation

Intraspecific Body Size Variation and Its Origins

Before going on to consider evidence for intrapopulational variation in mating success in digger wasps, we must pause to examine the patterns and origin of intraspecific variation in body size among males. Such variation can be considerable (Fig. 3-6). The largest males of *crabroniformis* we collected had a mass 6.6 times greater than the smallest males in the population. Equivalent values for *psyche* and *basilaris* are at least 3.2 times and 3.7 times, respectively. Table 9-5 summarizes intraspecific variation in head width in seven species of beewolves (O'Neill, 1983a, 1983b, 1985; O'Neill and Evans, 1983a).

Much intraspecific variation in the body size of adult aculeates can be explained by variation in the provisions they received as larvae (see review by O'Neill, 1985). In *Trypargilum politum*, variation in calories provided per cell explains 94% of variation in adult size (Cross et al., 1978). Significant correlations between adult size and amount of provisions have also been found in other solitary bees and wasps (Alcock, 1979; Cowan, 1981; Freeman, 1981; Jayasingh and Taffe, 1983). We are not aware of any studies demonstrating a lack of such a correlation. Variation in the size of eggs produced by females (O'Neill, 1985) could also contribute to variation in final offspring size (Jayasingh, 1980).

The knowledge that variation in the amount of food provided to a developing digger wasp larva contributes strongly to its size as an adult still leaves us with the problem of explaining the origin of variation in the amount of provision. While genetic variation in provisioning ability may contribute to size variation, several environmental influences on the calories available to a larva seem particularly important in digger wasps and other aculeates with advanced female parental care (Alcock, 1979). Factors that may influence the effective amount of food provided to each offspring include the following:

(1) *Activities of cleptoparasites and fungi, which may steal or destroy*

Table 9-5. Variation in head width in males of 7 species of Philanthus

Species	Mean head width and SD (mm)	Range (mm)	N
basilaris	3.21 (0.21)	2.2-3.7	380
bicinctus	4.39 (0.29)	3.7-5.2	58
crabroniformis	3.13 (0.29)	2.4-3.8	63
psyche	2.02 (0.18)	1.6-2.3	124
pulcher	2.33 (0.22)	1.6-2.8	262
serrulatae	2.23 (0.17)	1.9-2.5	50
zebratus	3.20 (0.25)	2.4-4.1	258

part of the provision provided. Ants, for example, sometimes steal prey temporarily left near the entrance of the nest by female *pulcher* returning from foraging flights. In some species, including the philanthine *Clypeadon laticinctus*, conspecific females steal prey from one another (Alexander, 1986, and references therein).

(2) *Variation in nutritional quality of individual prey items.* Since larvae do not consume the exoskeletons of prey, variation in the ratio of hard parts to soft parts could influence the percent calories available from each prey item (McClintock, 1986). Female *zebratus* prey on bees and wasps of a variety of shapes, for example. It seems reasonable to conclude that a short, stout prey item (e.g., a bee of the genus *Osmia*) would be of greater value than an elongate item of the same mass (e.g., an ichneumonid wasp) because of differences in the percent body mass devoted to exoskeleton. Females may not be able to precisely evaluate such variation in prey quality. Their ability to monitor the total mass of prey placed in each cell may also be limited. Finally, smaller prey items may die sooner and desiccate more rapidly than larger ones, which in some cases remain paralyzed for 8–10 days (Lane et al., 1986).

(3) *Influence of temperature on development.* Soil temperature profiles at cell depth are not steep, and variations in soil temperature are much less pronounced than those in air temperature. In spite of this, cell temperature should vary somewhat with depth in the soil (Campbell, 1977), local conditions (e.g., percent vegetation cover), and the time of the season at which the cell is provisioned. The temperature at which an insect larva develops influences its final adult size (Maksimovic, 1958; Mackey, 1977; Scriber and Lederhouse, 1983), possibly because individuals developing at higher temperatures have a higher rate of respiration and, consequently, a lower rate of mass accumulation (Scriber and Lederhouse, 1983).

(4) *Temporal and spatial variation in availability and quality of resources.* The amount of provisions placed in each cell should vary with spatial and temporal fluctuations in resource availability. Female digger wasps may place prey of various sizes in each cell. In *pulcher*, for instance, the range in prey weights found in a given cell was as high as fivefold. Such variation could be the result of chance factors involving the size of prey encountered on given foraging trips. Temporal variation in availability and quality of prey can take several forms. Seasonal variation may influence the species of prey available on any given day. Thus, the time of year at which a female emerges can influence her foraging success (Hastings, 1986). In addition, daily weather fluctuations can influence the total length of the foraging period or the quality and quantity of prey available. Because *Philanthus* females probably

provision one cell each day, this variation could affect the total amount of provisions for each larva.

(5) *Intrinsic differences among provisioning females*. A female's size and possibly her age should profoundly influence her allocation of resources to offspring (Alcock, 1979). The effect of size is well documented in the Philanthinae. Within species, there are significant correlations between a female's body size and both the size of the prey she provisions with and the size of the eggs she produces (Linsley and MacSwain, 1956; O'Neill, 1985). In *Clypeadon laticinctus*, female body size correlates with the rate of provisioning (Alexander, 1985). In *Cerceris arenaria*, larger females can make a greater number of foraging trips per day because of size differences that affect heat exchange and body temperature (Willmer, 1985). A review of size influences is found in O'Neill (1985). Though little studied, age may also correlate with provisioning ability, as older female *Philanthus* have tattered wings and reduced fat stores.

Some factors mentioned above are speculative to a greater or lesser extent, and some lack strong empirical support, yet all represent logical possibilities derived from our knowledge of the biology of digger wasps and other insects. Overall, we envision the environmental factors working in the following manner. Factors 1–4 and probably others act together to generate differences in body size among males and females. Many of these factors may be random with respect to genetic differences among females. Size differences among females, amplified by the well-known effects of female body size discussed under Factor 5, result in even greater variation in resource allocation to individual offspring. Essentially, we are postulating that variance in size has a lower limit dictated by stochastic (or at least unpredictable) influences in the environment, possibly combined with recurrent mutations that increase variance. Natural selection would favor the spread of any genes that countered such effects, so variation may attain some relatively stable value. Consistency in size variation between years is apparent in our studies of *Philanthus* and is well documented in the bee *Centris pallida* (Alcock, 1984). Comprehensive experimental studies of the heritability and the specific environmental correlates of size variation are needed.

Intraspecific Variation in Mating Success among Males

Variation in body size among males in populations of digger wasps strongly influences the males' reproductive success. Although no studies have traced the effect of body size all the way to a direct measure of male reproductive success, such as the number of adult offspring, less direct measures show the relationships outlined in Table 9-6 (see Tables 9-1 and 9-2 for references).

As a result of (1) the relationship of size to success documented in Table 9-6, (2) limited numbers of territory sites in some species, and (3) possible limitations on the production and storage of pheromone, large numbers of male beewolves are excluded from territories. As evidenced from the removal experiments and territory censuses we conducted for *basilaris*, *crabroniformis*, *psyche*, and *pulcher*, not only are these males usually smaller than those that maintain residency, but they represent a majority of the males active in a population. In the latter three species, for example, sequential removals of males from territories showed that, on average, there were from 1.4 to 2.4 replacers for every original resident removed. That is one indication that the level of competition is high in beewolf populations.

Assuming that most, if not all, matings occur on territories, what options are available to this multitude of small males that might afford them some chance of mating? Small males generally lose fights on territories. In fights involving *basilaris* males below average in body size, for example, only one such male was able to usurp a territory ($N = 17$), and only one successful defense by such a male was recorded ($N = 9$). Because they remain in the area and persist in their search, however, these small males are sometimes able to occupy and scent-mark perches. Our observations show that they can come upon possession of a territory in two ways. First, they have some chance of finding a

Table 9-6. Importance of male body size in digger wasps (Sphecidae)

Male body size correlates with	Species in which relationship is documented
Mating success	Philanthus zebratus, Bembecinus quinquespinosus
Number of territories occupied	P. basilaris, P. crabroniformis, P. psyche, P. pulcher
Success in aggressive interactions	P. basilaris, P. crabroniformis, P. pulcher, P. zebratus, Eucerceris flavocincta, Lindenius columbianus, Trypargilum tenoctitlan
Territory quality	P. bicinctus, P. basilaris
Ability to avoid predation while on territory	P. basilaris

territory defended by an even smaller male. Although male *pulcher* in the lower third of the size range for the species never won fights against larger males, for example, they were able to win 7 of 38 (18%) encounters on territories, including 3 in which they usurped a territory. Second, they can come upon an empty territory. Some such territories may have been recently abandoned and may still contain scent-marked plants. Our removal experiments demonstrated how swiftly such abandoned territories are occupied by roving males, who are generally smaller than the original residents (94% of 143 observations in four species of beewolves). Natural analogs of such removals exist when territory residents are preyed upon, leave to feed or mate, or leave for reasons not apparent to us. Frequently, small replacers have to continue their search a short time later when they lose a fight and have to leave the territory. Thus, in male beewolves we may define a single tactic: enter potential territories until a suitable one is found. The major difference among males may simply be a varying definition of *suitable*. For a very large male, *suitable* may mean the highest-quality territory available. For a very small male, it may mean the first unoccupied (and preferably scent-marked) territory available.

If most females were receptive only early in the season and if smaller males emerged before large males, then perhaps the advantages of large size would be offset by the relative absence of larger males when most females were available. Data from digger wasps suggests that there is no correlation between body size and emergence date within species. We have observed males of a wide size range appearing on the first several days of the season in a number of beewolf species (e.g., *pulcher* and *zebratus*). In any case, most males emerge before the peak of female emergence. More direct evidence exists for the digger wasp *Bembecinus quinquespinosus*, which we have caught in emergence traps. We found no correlation between emergence date and body size for either sex of this species (O'Neill and Evans, unpublished).

Although we have evidence to support the hypothesis that large males have a higher mating success than small males, the paucity of observed matings makes it difficult to quantify variance in lifetime mating success among individuals or phenotypes. Male beewolves almost assuredly have some potential for polygyny, given the disproportionate territorial success of large males. In addition, marked males are commonly observed patrolling or holding territories soon after they have been observed mating. In fact, males of *basilaris, bicinctus, crabroniformis, pulcher*, and *triangulum* have been seen flying directly back to their territories after having broken contact with a female they have just mated with. Apparently no delay is necessary before a male is ready to mate again. The longevity of some males certainly would contribute to a potential for polygyny. One large male of *basilaris*, for example, was observed holding territories on 20 days over a total period of 25 days in 1980. Unfortunately, we have only one record of a male beewolf (of *zebratus*) mating twice.

Intraspecific Variation in Male Mating Tactics in Digger Wasps

The males of a population may have access to more than one tactic for obtaining mates. There is a considerable body of literature dealing with alternative mating tactics in insects (see Thornhill and Alcock, 1983) and with the theoretical aspects of such variation (see Austad and Howard, 1984). We will follow Austad (1984, p. 310) in defining alternative mating tactics as "*any discontinuous variation*" in the form of mate-accruing behavior among males of "*a single population with associated differences in the behaviors' costs and/or benefits.*" We will not attempt an overview of the evolutionary models developed to explain alternative mating tactics. Rather, we will present specific examples from the digger wasps and discuss the adaptive nature of such behavioral variation.

Several factors favor the evolution of behavioral variation among mate-seeking males. First, variation in the competitive abilities of males may make it essentially impossible for certain individuals to succeed in obtaining a mate where most receptive females are located (e.g., on territories). Therefore, it may be better for them to search elsewhere, even if their likelihood of success is low. In an oft-quoted phrase, Dawkins (1980, p. 344) refers to this as "making the best of a bad job." In other words, a male switches tactics because a low likelihood of success with a secondary tactic is better than no likelihood of success with the primary tactic. Second, receptive females may be distributed in ways that require males to use completely different tactics in order to find and mate with them. These two factors are obviously related, since alternative opportunities available to less competitive males depend on the existence of variation in the availability of females.

As shown above, within species of digger wasps, large males often have greater mating success because of a superior ability to compete for females or for locations likely to attract them. Although most females mate with these large males, a small number may not and may thus be available to males undertaking a different tactic. Thus natural selection should favor behavioral flexibility so that small males undertake a tactic giving them access to the small pool of females not mated by the large males. This alternative would be an example of a condition-dependent tactic (e.g., Dominey, 1984). Males respond to a condition, in this case their size-related likelihood of success, by adjusting their behavior to maximize the probability of success. Several examples from digger wasps may fit this model.

In one of the Jackson Hole populations of *zebratus*, two very different male mating tactics are discernible (O'Neill and Evans, 1983a). In one group, males patrol above the nesting area, where they intercept females making flights to and from nests. In a second group, males establish scent-marked territories just outside the nesting area, in the manner typical for the genus. It appears that most females are mated by the former group, although further data in support of this assertion are needed. Males undertaking the patrolling tactic are larger on average than territorial males, even though patrolling is a nonaggressive tactic. Several males were observed to use both tactics, so it appears the behavioral variation is not the result of a genetic polymorphism. At present, although we do not know why small males tend to be excluded from the patrolling strategy, this population appears to offer an example of a size-dependent alternative mating tactic.

In another digger wasp, *Bembecinus quinquespinosus* (Nyssoninae), the reasons for the correlation between body size and the tactic undertaken are better understood (O'Neill and Evans, 1983b and unpublished). Females in this species nest in extremely dense aggregations (Evans et al., 1986), the offspring emerging in great numbers the following year over a period of 1–2 weeks. Some males in the population search and dig for females about to emerge from the ground, probably using odor as a cue to home in on them. Large numbers of males may be attracted to these digging sites, so that when a female finally appears, up to 50 males may be waiting for her. The resulting intense battle may last for nearly a half-hour. Large males are particularly likely to win such encounters, because they are better able to carry the female in flight away from the emergence area to avoid further harassment. The largest males, representing about one-quarter of the population, obtained over 90% of the matings in the emergence area. We also estimate that 80–90% of the females were mated before leaving the emergence area. No males below average in size were ever observed to mate there. Because some of the females emerge and leave the immediate vicinity without mating, however, a pool of virgins is available to small, less competitive males. These males patrol nonaggressively just outside the emergence area and intercept passing females, occasionally mating.

Following Alcock et al. (1978) we offer an admittedly oversimplified model to explain why behavioral variation persists in the two species just mentioned even though males undertaking one tactic probably have a much higher mating success on average than males undertaking the alternative tactic. The source of size variation and the size-dependence of competitive ability may be the keys. Consider a *hypothetical* ancestral population of a digger wasp in which there are three discrete and heritable strategies: some males patrol the nesting area, some establish territories outside of it, and some have the flexibility to do either. Suppose also that males exhibiting the patrolling tactic have the highest average mating success and that large males succeed best at this tactic, whereas smaller males consistently fail at it. Finally, assume that, as argued above, unpredictable environmental influences affect intraspecific size variation. If so, males with genes controlling the expression of each behavior type will be randomly distributed across size classes to some extent. Obligate patrollers will have high mean fitness if they happen to be large and essentially zero fitness if they are small. Obligate territorial males should expect low fitness (but non-zero fitness) whether they are large or small. Males capable of matching their behavior to their size-related probability of success will, on average, have high fitness if large (as patrollers) and low, but non-zero, fitness if small (as territorial males). Therefore, *on average*, males with flexibility should have higher expected mating success than males restricted to a single mating tactic, as they can match their behavior to their size-related abilities (recall that small obligate patrollers have essentially no mating success). Of course, this argument assumes that no major disproportionate costs are associated with maintaining flexibility. If our assumptions approximate reality and if our model is robust enough, males with behavioral plasticity should be most successful in the face of unpredictable environmental influences on body size, so this trait should spread through the population.

In some species, males may engage in opportunistic tactics while waiting for a chance to switch to a more profitable tactic. While moving about searching for available territories, for example, males may also have other opportunities for mating. In *basilaris*, *crabroniformis*, and *pulcher*, we have seen such intruders attempt to disrupt mating pairs, although none were successful. In *pulcher*, one intruding male was able to copulate with a female that had just been attracted by the resident.

We commonly observed nonterritorial male beewolves in the vicinity of active territories. In *Oxybelus sericeus* and *Trypargilum tenoctitlan*, males searching for single nests to guard will also attempt copulation with any females they come across. The overall significance of such "side-payment tactics" (Dunbar, 1982, p. 413) in digger wasps is not clear.

Variation in male behavior that is apparently not size-dependent but that exploits variable patterns of female availability has also been discovered. Males of *pulcher* patrol the nest site and attempt copulations with digging females before leaving to set up territories later in the morning. Some males of *bicinctus* establish territories in the dense nesting area, whereas others establish territories in an area containing flowers (nectar and hunting sites) and scattered nests. Early in the season, *Bembix rostrata* males search and dig for females in the emergence area, much in the manner of males of *Bembecinus quinquespinosus*. Later they attempt matings with females foraging on flowers or digging at their nests. Brief reports of other species (*Oxybelus bipunctatus* and *Cerceris truncata*) indicate that some males patrol the nest site while others patrol flowers. We expect that a more systematic search for alternative mating tactics in digger wasps would reveal a variety of such behavior patterns.

Costs Associated with Mating Tactics

A male adopting a particular mating tactic may also be exposing himself to a variety of risks. One such risk documented repeatedly in beewolves is predation. Males approach almost all insects that enter their territories. Although this behavior increases their chances of intercepting females and quickly repelling intruding males, it also brings them to the attention of predators (Gwynne and O'Neill, 1980). Territorial males of *basilaris*, *bicinctus*, *psyche*, and *pulcher* are preyed on by robber flies, large and formidable predators of the family Asilidae (Fig 2-10). Male *basilaris* and *bicinctus* are also preyed on by females of *basilaris* (O'Neill and Evans, 1981). Stubblefield (1985) reports that territorial males of *parkeri*, the smallest North American species, temporarily abandon their territories when female *basilaris* or robber flies appear. There is some evidence that larger territorial males of *basilaris* are better able to escape predators (O'Neill, 1983b).

Male digger wasps that patrol for females are also subject to predation. Ant-lion larva prey on males of *Plenoculus boregensis* (Rubink and

O'Neill, 1980), and *Bembix niponica* (Tsuneki, 1956) searching for females in nesting areas. Mate-searching males of *Bembecinus quinquespinosus* are preyed on by horned lizards, *Phrynosoma douglassi* (O'Neill and Evans, personal observations). It is not clear how important predation has been as a selection pressure influencing the evolution of mating strategies.

By searching or waiting for females in particular locations, males place themselves in a particular thermal environment. Territorial males of *Philanthus* are often active in midsummer, occupying arid, fully insolated areas. In fact, they generally abandon territories if shade encroaches or if clouds obscure the sun for very long. We saw this behavior consistently in *crabroniformis* and *pulcher*. For *psyche*, surface temperatures may reach above 60°C within territories. Thus, males perching on the ground within territories place themselves under considerable thermal stress. We have already noted that if forced to remain on the ground at such temperatures, they may die in a matter of seconds (O'Neill and O'Neill, in press). Several options are available to males territorial in such environments. First, they can leave the territory and search for females elsewhere. We have already argued however, that for male *psyche* the nest site is the place where females are most likely to be encountered. Second, they can take advantage of microclimatic variation within their territories. They do so by perching on the ground for shorter periods later in the day as temperature increases and by switching to perches on plants. Perching on plants places them in an area with lower temperatures and removes them from the still air of the boundary layer close to the ground. Males of other species occupying thermally stressful environments (e.g., *barbiger* in the San Rafael Desert, *tarsatus*, and *multimaculatus*) also tend to perch on plants.

Variation in color among species of beewolves may also be related to the thermal environment frequented by males. Males in species that are territorial in thermally stressful environments, such as *multimaculatus*, *parkeri*, and *psyche*, are more extensively yellow than males in less stressful locales (e.g., *bilunatus*, *pulcher*, and *sanbornii*). Males of *albopilosus*, which search for females in open dunes, have more extensive yellow coloration in the southern part of their range (Texas) than in the north (southern Alberta). Although body coloration can influence body temperature (Willmer and Unwin, 1981), the significance of species differences and geographic patterns of color variation among beewolves and other digger wasps remains to be fully investigated.

Intersexual Selection

Any answer to the question of whether female digger wasps choose among potential mates would be speculative. Intersexual selection is a complex subject (Bateson, 1983) that has not been directly addressed in any studies of digger wasps. Following Halliday (1983), we define female mate choice as any behavior pattern shown by females that makes them more likely to mate with certain males than with others. One major problem is that we have little idea how the characteristics of conspecific males vary in a way that influences the fitness of females mating with them. A second difficulty concerns the mechanisms by which females might evaluate male quality. Third, given that females can distinguish among males, can they exercise a choice, or do males control who mates with whom?

Excluding the minority of species of Sphecidae in which males contribute to the nesting success of females (e.g., *Oxybelus* and *Trypargilum*), male sphecids offer a female no material benefits that affect her fitness (i.e., nuptial gifts, food for young, access to a territory in which important resources occur, spermatophores, or assistance in nest building and maintenance). Beewolves serve as a prime example. The contact between a female and her mate generally lasts no longer than the 5 minutes or so necessary to transfer sperm (it may take only several seconds in some species of digger wasp; e.g., O'Neill and Evans, 1983b). It seems unlikely that anything besides sperm is given to the female. Upon visiting her mate's territory, the female does not gain access to important hunting sites, nectar sources, or nesting substrate. Either these resources are available elsewhere in abundance, or the male does not control the female's ability to use them. In theory, it is possible that females' choice of males is based on variation in male genetic quality, a form of intersexual selection that is difficult to study (Boake, 1986). That leaves the problem of how females could evaluate genetic differences. However, because male Hymenoptera are haploid, recessive as well as dominant traits are expressed (Brückner, 1978).

Given these patterns and difficulties, can we discern any form of female choice in *Philanthus*? At best, we can say that mating is not random in species with territoriality. By virtue of their large size, females usually successfully refuse males attempting to mate at locations other than territories. Even when their nests occur within territories, females are able to repel potential suitors (e.g., in *bicinctus* and *psyche*). The females fly to scent-marked territories and mate with the resident males, so they are exercising some form of choice. These males have undertaken behavior that has allowed them to successfully defend their territories, have scent-marked, and tend to be above average in body size; but we do not know the genetic correlates of these traits. In fact, as we argued earlier, body size may have low heritability. If genes influencing male abilities do vary, it may be that the combination of intense intrasexual selection and the fact that females choose to mate at territories makes it probable that a female will mate with a male of above average "quality." Any additional female behavior that enhances assessment of males may be risky and time consuming relative to the benefit to be gained (O'Neill, 1983b). Given the multitude of uncertainties expressed here, we must leave further analysis of this phenomenon until more data are available.

Problems for Further Study

This chapter has discussed a variety of factors that may influence the evolution of male mating strategies in beewolves (see also Fig. 10-2). It is clear from this review that our knowledge of male behavior in digger wasps in general remains a patchwork. In consequence, our evolutionary scenarios and adaptive hypotheses should be considered simply a first step to a more complete synthesis; however, we believe that the substantial increase in data on male sphecids that has been accumulated over the last 10–15 years provides a satisfying beginning. Beewolves have played a major role in this advance and furnish excellent systems for future studies. Of particular value would be more direct information in the following areas: (1) patterns of female receptivity; (2) the relationship between temporal and spatial distribution patterns of mate-searching males and receptive females (including estimates of the operational sex ratio); (3) lifetime mating success of individual males or males of particular phenotypes; (4) the dynamics, function, and limitations of scent marking by male philanthines; (5) the degree to which predation risk and the physical environment constrain the evolution of mating strategies; (6) the role of female choice in digger wasp mating systems; (7) the occurrence, form, and payoffs associated with alternative mating tactics; and (8) the relationship of environmental influences on body size to the evolution of alternative mating tactics. Although the comparative method will continue to play a major role in such studies, we hope an increased number of experimental and quantitative observational methods will be used as well.

Chapter 10

Major Features of Nesting Behavior, with a Final Look at Beewolves

In this final chapter we first review some of the more salient features of female nesting behavior. With this and the preceding chapter as a basis, we then stand back and attempt a broad view of beewolves as products of evolution: the apparent adaptiveness of their behavior patterns and the significance of the diversity within the genus. There may be few final answers, but there will be food for thought and ideas for further research on these abundant and attractive insects.

Nesting and Predatory Behavior

In contrast to the quite recent flourishing of interest in the behavior of male digger wasps, reviewed in the preceding chapter, female behavior has intrigued observers for a very long time. Sphecid biologists are certainly not subject to the criticism, leveled at vertebrate ethologists, that research is unduly biased toward male behavior (Waser and Waterhouse, 1983). The literature on female behavior is voluminous. However, although there are several modern reviews (Malyshev, 1968; Evans and West-Eberhard, 1970; Iwata, 1976; Bohart and Menke, 1976), none of them has provided a wholly satisfactory perspective on the evolution of this complex behavior. Considering the size of the group, the many gaps in our knowledge, and the lability of behavior compared with structure, a better perspective is not likely to be forthcoming soon. Here, we discuss nesting behavior under several subheadings, hoping to provide at least a preliminary framework for understanding the position of *Philanthus* among the Sphecidae and for understanding trends in evolution within the genus.

Habitat and Nest Distribution

Although called digger wasps, Sphecidae nest in a diversity of habitats, including hollow twigs, rotten logs, and free mud nests well above the ground. All Philanthinae, however, are ground nesters, a feature widespread among higher Hymenoptera and probably ancestral within the family. Within the Philanthinae there is a major dichotomy. Cercerini make vertical burrows, which they dig primarily with their mandibles and clear with a pygidial plate at the end of the abdomen; they are *pushers*, in the terminology of Olberg (1959). Vertical burrows dug in this manner are unusual in Sphecidae, though common in bees; they appear associated with the ability to occupy soil that is relatively hard packed. In contrast, other Philanthinae (so far as studied) make an oblique burrow; these are *rakers* that use a series of spines on the front legs to scrape soil from the burrow.

Types of soil occupied by species of *Philanthus* range from the bare, uniformly grained sand of dunes and blowouts (*albopilosus, psyche*) to patches of coarse sandy gravel among vegetation (*bicinctus, barbatus*) or sandy loam in wooded areas (*pulcher, bilunatus*). Analyses of the physical factors influencing choice of a nesting site, similar to that of Brockmann (1979) on *Sphex ichneumoneus*, have not been made for philanthine wasps, but distinct differences would surely be found among the species of *Philanthus*.

It is not unusual to find two or more species of beewolves occurring in the same or closely similar sites, either sequentially in the course of a season (e.g., *pulcher* and *bilunatus*, Chapters 6 and 7) or simultaneously (e.g., *gibbosus* and *crabroniformis*, Alcock, 1974). O'Neill and Evans (1982) found four species occurring at the same time along a gulley in north central Colorado and showing slight to moderate overlap in nest sites.

The richest site known to us for *Philanthus* species is the San Rafael Desert, Emery County, Utah. In a 3-day period, 6–8 August 1981, we collected 10 species of beewolves along about 15 km of roads north and east of Goblin Valley State Park. Although we found only a few nests and male territories over the 3-day period (cited in earlier chapters), all species appeared to occupy similar, sandy desert habitat. Frank D. Parker of the U.S. Department of Agriculture Bee Biology and Systematics Laboratory, Utah State University, has collected three additional species in this area. Some are most abundant in June (*pulcher, serrulatae*), others in midsummer (*psyche, multimaculatus*), others in August and September (*basilaris, barbiger*) (F. D. Parker, personal communication, 1986).

Other Philanthinae, especially *Clypeadon laticinctus* and *Aphilanthops frigidus*, commonly nest in the same sites as *Philanthus* species. Indeed, these sites are often shared with a variety of ground-nesting wasps of several subfamilies of Sphecidae and of the families Pompilidae and Vespidae (Eumeninae).

The extent and patchiness of soil of suitable quality influence local population size and density not only in *Philanthus* but in many groups of wasps. Although we have classified species such as *zebratus* and *pulcher* as having the highest nest density of those we have studied (Chapter 9), we have also reported that populations of both species at less favorable sites have more scattered nests. In the broad expanses of sandy soil adjacent to the Great Sand Dunes of southern Colorado, *basilaris* nests are widely scattered (Fig. 3-8). In contrast, in a restricted area of eroded soil along a gully in northern Colorado, nests are more closely spaced (Chapter 3). Nevertheless, it is possible to categorize most species as to relative nest density under favorable conditions, as we did in Chapter 9 and in Tables 9-1 and 9-2 with respect to the influence of nest density on male reproductive behavior.

When nests are closely packed there may be considerable aggression among females, resulting in wasted time and energy (see especially *sanbornii*, Chapter 3, and *pulcher*, Chapter 5). We have noted no prey stealing at nest entrances comparable to that occurring among some species of *Bembix* (Nyssoninae; Evans, 1966c); however, Alexander (1986) has found that some females of the philanthine wasp *Clypeadon laticinctus* regularly enter nests of other females and remove prey to their own nests.

Nesting close beside other individuals may offer compensating advantages. In at least one sphecid wasp, *Crabro cribellifer* (Crabroninae), it has been shown that there is a positive correlation between internest distance and cell parasitism by *Metopia* flies (Wcislo, 1984). This finding has been interpreted as confirming the "selfish herd" hypothesis, which, for digger wasps, predicts that the probability that any nest will be attacked decreases as the number of active nests per unit area increases. Wcislo suggests that the presence of accessory burrows may further enhance the effectiveness of this behavior by increasing the number of apparent nearby nest entrances.

Once established at a suitable site, *Philanthus* populations often remain fairly stable in size for many years. The Yellowstone population of *bicinctus* is known to have occupied the same site for 30 years, the Great Sand Dunes population for at least 9 years. A population of *zebratus* in Grand Teton National Park has been active at the same site for at least 23 years. In each case, population size has remained reasonably constant from year to year. An aggregation of *gibbosus* at Bedford, Massachusetts, remained at 32–40 nests for 6 years; when an area immediately adjacent was cleared by bulldozing, several of the females moved into this area, but only a short distance (Evans, 1974).

There are many other records of beewolves persisting at the same site year after year. On the other hand, there are records of the disappearance of the wasps from some sites. Tinbergen (1958) noted a marked decline in the numbers of *triangulum* nesting in the dunes of Hulshorst, Holland, causing him to terminate his studies there. Simonthomas and Simonthomas (1972) noted an abrupt decline in populations of this same species at sites in southern France, possibly the result of the attacks of cuckoo wasps. In discussing the North American species *serrulatae* in Chapter 6, we noted a decline in a population at the Pawnee National Grasslands, Colorado, from 79 nests in 1982 to 18 nests in 1983. In 1984 only one nest could be found, and in 1985 none at all; the wasps had also completely disappeared from flowers they had formerly visited in great numbers. Sites where the wasps nested in 1982 were only slightly more overgrown with grasses, and at no time did we notice an abundance of miltogrammine flies, cuckoo wasps, or mutillids. Possibly the lack of rainfall at critical periods rendered the soil too dry and hard for nesting.

Although population declines are probably sometimes caused by natural enemies, climatic factors and changes in vegetation and soil quality appear to be more often involved. When physical factors remain fairly constant, beewolf populations can be relied on to remain in place in roughly the same numbers years after year. This consistent concentration of nests at restricted sites would seem to provide an unusual opportunity for the buildup of predators and parasites, and many types have been recorded (Table 10-5). That the wasps are able to flourish in the face of such pressure from natural enemies we believe to be the result of several aspects of female behavior that have evolved in this context. That is a theme to which we return repeatedly in this chapter.

Nest Building and Maintenance

In a general way, the digging behavior of beewolves resembles that of many groups of digger wasps, soil being loosened with the mandibles and scraped from the burrow with synchronous strokes of the front legs. A mound of soil accumulates outside the entrance, often crossed by a series of grooves formed by the female as she backs from the burrow scraping soil. In two other genera of Philanthinae, *Aphilanthops* and *Clypeadon*, females spray soil a considerable distance from the entrance, up to 20 cm; thus no distinct mound of soil accumulates (Evans, 1962a). In *Cerceris* and *Eucerceris*, which make vertical burrows by pushing up soil with their pygidial plates, a circular rim of soil is formed around the entrance (Olberg, 1959). Both these modes of digging, while not unique in the Sphecidae, are sufficiently uncommon to be safely categorized as more advanced than that in *Philanthus*. (See Table 1-1 for the presumed relationships of these genera.)

Leveling Behavior. In four of the species of *Philanthus* we have studied (Table 10-1), the mound at the entrance is leveled fairly completely by a stereotyped series of movements performed either before or after the initial closure is made (described in detail under *politus*, Chapter 6). Evidence that cuckoo wasps (Chrysididae) as well as hole-searching miltogrammine flies (such as *Phrosinella*) are attracted to mounds at nest entrances seems to support a hypothesis that this behavior has evolved in response to these natural enemies. When leveling is combined with nest closure, entrances are virtually invisible to a human observer unless the female is seen entering or leaving. On the other hand, the large, spreading mounds of species such as *bicinctus* and *zebratus* can readily be spotted from a distance. In these large species, which bring prey from a considerable distance at a height of a meter or more, the importance of the nest mound as an orientation cue may override its importance as a cue to natural enemies. The ability of mound levelers to find their nests readily seems quite remarkable, especially since some of them (*pulcher*, in particular) nest in dense aggregations.

Leveling behavior has not been observed in other Philanthinae and must be considered a derived feature of beewolf behavior. As mentioned above, species of *Aphilanthops* and *Clypeadon* spray the soil widely; thus leveling would serve no function. Both behaviors—leveling and spraying—occur elsewhere in the Sphecidae primarily among the more advanced genera of Nyssoninae, spraying in *Bicyrtes* and *Microbembex* and leveling in several species of *Bembix* (Evans, 1966c).

Temporary Closure. In the majority of digger wasps, females close the entrance upon leaving and reopen it with several strokes of the front

Table 10-1. Summary of major features of nesting behavior in Philanthus (species arranged according to chapter in text)

Species	Nest type[1]	Nest closure	Accessory burrows	Leveling behavior
bicinctus	PS	+	+ or −	−
basilaris	PD	+	+ or −	−
zebratus	PD	+	+ or −	−
sanbornii	PD	−	−	−
gibbosus	PS	+	+ or −	−
crabroniformis	PS	+	+ or −	−
barbatus	PD	+	+	−
multimaculatus	PS	+ or −	−	−
inversus	PS	−	−	−
pulcher	DD	+	−	+
pacificus	DD	+	−	+
barbiger	DD	+	+ or −	−
psyche	DD	+	+ or −	+
serrulatae	DD	+	+ or −	−
politus	DD	+	−	+
ventilabris	PS	+	+	−
lepidus	PD	+	+	−
bilunatus	PD	+	+	−
solivagus	PS	−	−	−
albopilosus	DD	+	+	partial
triangulum	PS	+ or −	−	−

1. Nest types: DD= declinate, diffuse cell pattern; PD= proclinate, diffuse cell pattern; PS= proclinate, serial cell pattern.

legs upon returning; they also close from the inside while working or resting in the nest. These statements apply to most *Philanthus* species (Table 10-1). However, females of three well-studied species (*sanbornii*, *solivagus*, and *inversus*) consistently omit the closure while hunting, and variation has been noted in two others (*multimaculatus* and *triangulum*). One assumes that closure serves to deter oviposition by bombyliid flies and entry by hole-searching miltogrammine flies and marauding ants. On the other hand, closure tends to delay entry by females bringing prey to the nest and may thus enhance the success of satellite flies. The relative abundance of these natural enemies over time may have influenced the evolution of closing behavior among species and among local populations of some species.

Iwata (1976) compiled data on nest closure in a great number of ground-nesting wasps. Within many genera of diverse subfamilies of

Sphecidae there are examples of species that normally close the nest entrance and others that do not. Iwata's tables tend to obscure the fact that intraspecific variation has occasionally been described (see, e.g., Evans, 1966c; Kurczewski, 1982; and examples in Table 10-1). In general, however, closure or nonclosure of the nest entrance while the female is hunting can be considered species-specific or in some cases characteristic of genera. None of the many species of the genus *Cerceris* that have been studied make such a closure, for example, but in some species of this genus communal nesting, involving the presence of a guard that plugs the nest entrance, has evolved (Evans and Hook, 1982a, 1982b).

Permanent Closure. After a cell has been fully provisioned and the egg laid, the cell burrow is tightly packed with soil. The main burrow may be packed with soil after a full complement of cells has been prepared. In *Philanthus*, the latter practice has only rarely been reported in species making proclinate, long-term nests, and it is probable that in these species females continue to add cells to an active nest until they die (it is not unusual, late in the season, to find a dead female in such a nest). On the other hand, in species making declinate, short-term nests, it is common to see females completely filling the main burrow before starting a new one (see under *pulcher*, Chapter 5). In the majority of digger wasps of diverse groups, as well as in spider wasps (Pompilidae), permanent closures are a fixed part of the behavioral repertoire, as females commonly make a series of short-term nests.

Accessory Burrows. In *Philanthus*, accessory burrows are of irregular occurrence; they are consistently present in only a few species and consistently absent in only a few (Table 10-1). When intraspecific variation occurs it is usually intrapopulational. Accessory burrows evidently sometimes furnish fill for closure and thus might be expected to occur in species nesting in firm soil, which provides little loose soil for fill (e.g., in *gibbosus*). Although accessory burrows are generally absent in some species occurring in sand (*psyche*), these burrows are a consistent feature of certain species occurring in very friable sand (*albopilosus*, *lepidus*). In many cases accessory burrows are dug after the closure has been completed and thus do not serve as quarries, although they may have evolved from such quarries. Females sometimes rest in these burrows, and they frequently clear them out or add to them over time (see especially *lepidus* and *bilunatus*, Chapter 7). In species in which the true burrow is never closed, there would be little selection pressure for the development of accessory burrows. In fact they have not been observed in the three species that consistently omit closure (*sanbornii*, *inversus*, and *solivagus*).

Accessory burrows occur in digger wasps of several groups (though not, so far as known, in other genera of Philanthinae). There is evidence that they often divert the attention of hole-searching parasites and at times cause them to oviposit or larviposit at inappropriate sites (Evans, 1966a). Tsuneki (1963) found that the Asiatic species *Sphex argentatus* (Sphecinae) consistently makes accessory burrows and the sympatric species *S. flammitrichus* does not. Nests of *S. argentatus* showed cell parasitism by miltogrammine maggots varying from 0 to 21% at four different sites. At three sites *S. flammitrichus* cells showed parasitism varying from 33 to 44%. At the one site where the two occurred together, *S. argentatus* cells showed 9% parasitism compared with 44% in *S. flammitrichus* (see also Edmunds, 1974, Fig. 12.10).

Wcislo (1986) has shown that flies of the genus *Metopia* are attracted to artificial burrows in the soil. Both *Metopia* and bee flies (Bombyliidae) have been observed to larviposit and oviposit in empty burrows (Evans, 1966a; Endo, 1980).

In *Philanthus*, the correlation between the incidence of accessory burrows and the percentage of cells parasitized is by no means as striking as that Tsuneki found in *Sphex* (see below, under "The role of Natural Enemies in the Evolution of Behavior"). That accessory burrows have evolved primarily as a response to hole-seeking parasites nevertheless seems a reasonable hypothesis.

Nest Structure. There is much variation in nest structure among digger wasps as a whole, and several nest types occur in *Philanthus*, as outlined in Chapter 2 (see Table 10-1). As Iwata (1976) has pointed out, nest structure is often variable within major groups of digger wasps, and that is true of Philanthinae as well as other major subfamilies. Iwata has compiled data on the nests of many ground-nesting wasps, but unfortunately few consistent patterns are evident from the data.

Single-celled nests are usually considered ancestral among the Sphecidae (Evans and West-Eberhard, 1970; Iwata, 1976). Thus it might be concluded that, within *Philanthus*, short-term nests with few cells (as in members of the *pacificus* and *politus* groups) represent a type from which the longer-term, many-celled nests of other species have been derived. We doubt if that is the case, however. Declinate nests, in which vertical cell burrows are built from the end of a gently sloping oblique burrow (e.g., Fig. 6-6), are unusual among Sphecidae. Builders of such nests make only a few cells per nest and then move on to a new one, thus "putting few eggs in one basket." They may not only reduce the opportunities for natural enemies to move from cell to cell but may shift to a different site if the initial substrate becomes unsuitable.

In contrast to declinate nests, proclinate nests are widespread throughout the Sphecidae and so may represent the ancestral type in *Philanthus*. In Sphecidae as a whole, proclinate nests may be short term, with one or a few cells, or long term; but in *Philanthus* they are associated with long-term, many-celled nests.

There is much uncertainty concerning the dichotomy between progressive and regressive cell patterns. Although regressive nests appear characteristic of *sanbornii* (confirmed in two widely separated localities), variation in this feature has been reported in several other species. Only very careful nest dissections can reveal the exact sequence in which cells have been prepared, and it is best to await further data from *Philanthus* and from other groups of Sphecidae before drawing conclusions concerning the distribution and significance of this difference in behavior.

With respect to the dichotomy between diffuse and serial cell patterns, it seems clear that the latter is more advanced, as it is unique or nearly so in *Philanthus*. In soil that is friable to a considerable depth, a serial pattern would seem a practicable way to make cells from short cell burrows. However, in soil that contains many rocks or roots or that is more compact at increasing depth, a diffuse cell pattern may be more economical of time and effort. We have noted that although *gibbosus* builds cells serially in most substrates, when nesting in hard substrates it may use a diffuse cell pattern (Chapter 4).

Long-term, many-celled nests would seem especially susceptible to the attacks of Mutillidae and hole-searching flies. It is odd that none of the species making nests of this kind exhibits mound-leveling behavior, since mounds may serve as cues for natural enemies. The great depth of such nests and the firm closures of cell burrows doubtless act as deterrents. Also, the long burrows of such species often abruptly rise and fall, resulting in a serrate profile (e.g., Figs. 3-2, 3-9A, 3-16A, 7-3). Such nests are especially difficult to excavate, and it is perhaps not rash to assume that hole-searching natural enemies are retarded by the abrupt changes in the angle of the burrow.

Interpopulational variation in burrow length and cell depth has been reported for several species and is doubtless associated with differences in soil texture and moisture and with vegetational differences that

Major Features of Nesting Behavior   239

influence the structural integrity of the soil (see especially *gibbosus*, *pulcher*, and *albopilosus* as well as Tables 5-1, 6-1, 6-5, 7-3, and 7-7).

In summary, we believe that proclinate nests with diffuse cell patterns may represent the ancestral condition in *Philanthus*. Such nests occur widely in the Sphecidae, including other genera of Philanthinae (*Aphilanthops*, *Clypeadon*, *Cerceris*, and *Eucerceris*; the latter two genera make vertical burrows, but cell pattern is nevertheless diffuse). From this type may have evolved the serial cell patterns of several species and the short-term, declinate nests of several others. Among digger wasps, these two derived nest types are nearly unique to *Philanthus*.

Nest Reutilization and Nest Sharing

In a few species of beewolves, it is known that nests may be reoccupied and expanded by members of a second generation and that members of this generation may remain together in the nest for several days, sometimes longer. This behavior has been reported in the European *triangulum* (Chapter 8) and in the North American *ventilabris* (Chapter 7) and *gibbosus* (Chapter 4). In *gibbosus*, two females have occasionally been seen provisioning the same nest well along in the nesting season. It is noteworthy that all three of these species range well into subtropical areas. In *Cerceris*, where communal nesting is much more prevalent, it is chiefly characteristic of species occurring in warmer parts of the globe. There is also evidence of communal nesting in *Trachypus*, the neotropical ecological equivalent of *Philanthus* (Hook, 1985). Evidence of nest sharing should be looked for in other species of *Philanthus* making long-term nests and ranging into warmer parts of the continent (e.g., *multimaculatus*).

Predatory Behavior

General features of hunting and stinging behavior were reviewed in Chapter 2. Steiner (1986) has recently summarized comparative data on stinging patterns among solitary wasps. More advanced groups of Sphecidae, especially Nyssoninae and Philanthinae, tend to prey on more advanced groups of insects (especially Diptera and Hymenoptera), groups in which the nerve centers for locomotion are concentrated in the thorax. Many Sphecinae and Larrinae sting the prey (often Orthoptera or Lepidoptera larvae) several times along the underside of the body, but a single sting suffices in *Philanthus* and other more advanced Sphecidae. In most Pompilidae and in some groups of Sphecidae (many Sphecinae and Larrinae, for example) paralysis of the prey is incomplete or transient. In contrast, prey of *Philanthus* are deeply paralyzed.

The venom of *P. triangulum* differs from that of several other Sphecidae in containing acetylcholine but no histamine (Piek and Spanjer, 1986). Paralysis of honey bees by females of this species is caused by a combination of three low-molecular-weight components (called philanthotoxins), each of which has little or no effect alone. The venom acts both in the central nervous system and on neuromuscular transmission. Piek and Spanjer (1986, p. 288) remark that "it could be that some wasp species that prey on (for them) dangerous insects, such as other Hymenoptera, have developed peripherally acting toxins in addition to centrally active toxins." A broad-scale comparison of the toxins of Sphecidae has yet to be made.

Evans (1962b) recognized eight different ways in which wasps carry prey to their nests and arranged them in apparent phylogenetic sequence from most generalized (carriage backward over the ground with the mandibles) to most advanced (carriage on the posterior end of the body by structural modifications of the abdomen). All Philanthinae carry the prey in flight; there are examples of mandibular carriage (*Cerceris*, *Eucerceris*); pedal carriage, with the legs (*Philanthus*, *Aphilanthops*); and carriage on the tip of the abdomen (*Clypeadon*, *Listropygia*). Carriage of the prey beneath the body, primarily by the middle legs, is believed to protect the prey from satellite flies and to permit the wasp to open the nest entrance with the front legs while still covering the prey.

Approach to the Nest. In many species of *Philanthus*, females have been seen to exhibit devious flight patterns as they descend to the nest entrance. Females followed by satellite flies have often been seen to elude the flies successfully by undertaking freeze stops (Fig.10-1), circling behavior, or other tactics (see, e.g., under *crabroniformis*, Chapter 4, and *barbiger*, Chapter 5). Alcock (1975b) discussed this behavior and presented a diagram of apparent evolutionary pathways of approach flights. We present an outline (Table 10-2) somewhat expanded and modified from that of Alcock.

Why have beewolves diversified in this behavior rather than exhibiting similar patterns to serve the same function? Alcock (1975b) has suggested that diversity may enhance the effectiveness of the behavior

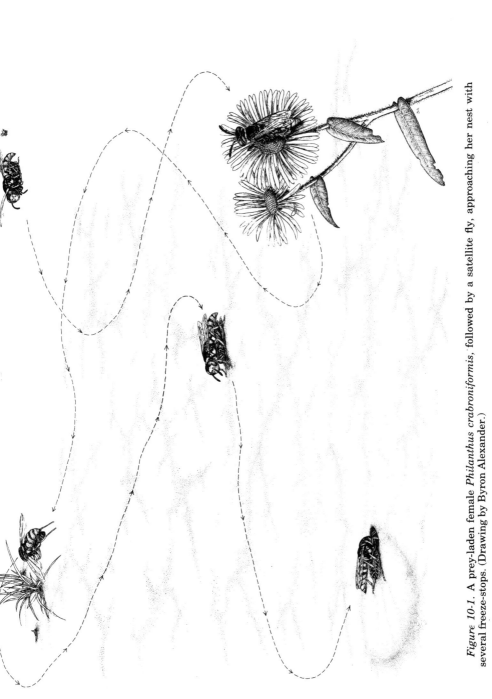

Figure 10-1. A prey-laden female *Philanthus crabroniformis*, followed by a satellite fly, approaching her nest with several freeze-stops. (Drawing by Byron Alexander.)

Table 10-2. Diversity of approach flights by female Philanthus

LOW APPROACH FLIGHTS (5-30 cm high)	HIGH APPROACH FLIGHTS (At least 0.5 m high)
With freeze stops	With freeze stops
crabroniformis multimaculatus politus bilunatus	zebratus inversus ventilabris
Departs, circles, and re-approaches in presence of satellite flies	Departs, circles, and re-approaches in presence of satellite flies
pulcher barbiger lepidus	basilaris solivagus
	Stepped descent with wavering flight
	gibbosus
Direct approach; no evasive flights noted	Direct approach; no evasive flights noted
pacificus psyche serrulatae albopilosus	bicinctus sanbornii barbatus

against parasites, particularly when two or more species occur in the same nesting area, as often happens. In a mixed aggregation of wasp species, the flies may be confronted with a variety of approach flights and may not be able to vary their behavior so as to successfully attack all of these species. Thus the probability of attack on individual females will be reduced.

The theory just described has been called the escape tactic diversity hypothesis (Schall and Pianka, 1980). The effectiveness of diversity in predator avoidance tactics among species occurring in the same habitat has only rarely been demonstrated, however. Schall and Pianka found that sympatric species of whiptail lizards differ in their escape behaviors and that these behaviors are more divergent than if each species, had evolved its escape tactics independently of others.

McCorquodale (1986) tested this hypothesis by examining the provisioning flights of six species of Sphecidae occurring together at a site in southern Alberta. Included were two species of *Philanthus, gibbosus* and *inversus*. The satellite fly was the ubiquitous *Senotainia trilineata*. He found that individual marked flies did not specialize on certain flight patterns but approached *Philanthus, Cerceris, Bembix,* or even epeoline bees regardless of their approach patterns. He concluded that diversity as such probably played no role in defense against the flies.

On the other hand, the flight patterns of the wasps after flies had detected them did significantly reduce the probability of contact. This effect was especially marked in the case of *Crabro argusinus* (Crabroninae), in which the use of a long, zigzagging approach flight reduced the frequency of contacts by more than 10-fold. Another interesting result of McCorquodale's studies is the observation that approach flights did not differ importantly whether or not a satellite fly was in pursuit. There was a tendency (in *P. inversus*) for followed flights to be less direct and to include more stops than nonfollowed flights, however.

Prey Type. Beewolves tend to prey on Hymenoptera that are abundant near their nests, enabling them in good weather to provision cells rapidly. Some, perhaps all, species have the capacity to take bees either on flowers or at their nests. At times, species nesting at the same site exploit much the same species of bees, taking advantage of those that are locally abundant. For example, Cazier and Mortenson (1965) recorded the same species of bees from nests of *gibbosus* and *multimaculatus* at sites near Portal, Arizona; and Alcock (1974) found that *gibbosus* and *crabroniformis* nesting along a path in Seattle, Washington, used virtually identical prey. These three species of beewolves are much the same size, and in fact all belong to the same species-group.

More commonly, when beewolf species occur in the same area, they tend to use different prey or to show differences in seasonality or nesting substrate. Evans (1970), for example, found that five species nesting in Jackson Hole, Wyoming, differed in size and consequently tended to use prey of different mean size. That was not true of the species pair *pulcher-crabroniformis*, which generally used the same small halictid bees; but *crabroniformis* began to nest at the time when *pulcher* populations were declining. *P. pacificus* also used much the same prey and nested at the same time as *crabroniformis*; but these two species nested in quite different types of soil, and their nests were consequently many meters apart.

O'Neill and Evans (1982) studied four species of beewolves nesting simultaneously in late summer along a gulley at Chimney Rock, Larimer County, Colorado. Again, the species were graded as to size, *barbiger* being the smallest (with a mean head width of 2.75 mm), followed by *basilaris* (mean head width, 3.62 mm), *inversus* (mean head width, 3.67 mm), and *bicinctus* (mean head width, 5.45 mm). The difference in head width between *basilaris* and *inversus* was not statistically significant, and they did use prey of similar size (Fig. 10-2); however, *inversus* was a specialist on male bees of the genus *Agapostemon*, which were used only occasionally by the generalist hunter *basilaris*. Thus there was limited taxonomic overlap in the prey of the two species. Because of their differences in size, *barbiger* and *bicinctus* showed no overlap in prey size and only small overlap with the other two species (Table 10-3).

It is tempting to conclude that differences in patterns of prey utilization among coexisting species may have arisen in a context of competi-

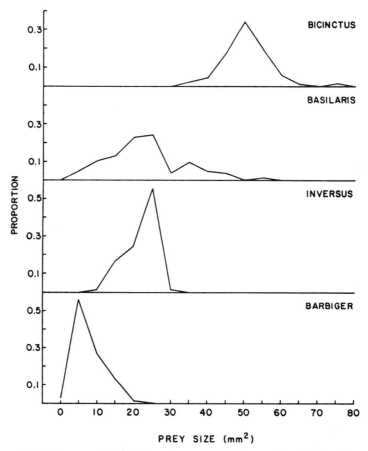

Figure 10-2. Frequency distributions of prey size for four species of *Philanthus* nesting at Chimney Rock, Larimer Co., Colo. Prey items were combined in five size categories. Size is calculated as thorax width times total body length. (From O'Neill and Evans, 1982, *Journal of Natural History*, courtesy of Taylor and Francis Ltd., London.)

tive exclusion. Although it is interesting to elucidate such patterns, there are difficulties in identifying the role that interspecific competition plays in determining patterns of species diversity and resource utilization (Schoener, 1974). Wasp size, hunting tactics, substrate preferences, and seasonality may all have evolved in other contexts and may incidentally enhance the ability of species to live in the same area. Because prey is carried in flight, each species must use prey within a certain size range; prey too bulky or heavy could not be carried, and prey below a certain size would require a great many foraging flights to provision a cell. There is little evidence that prey abundance is a limiting factor in female nesting success. Gwynne (1981), however, found that *bicinctus* females took longer to provision cells and made fewer cells per nest during a year of relative bumble bee scarcity, when they frequently used bees other than bumble bees and sphecids of the genus *Podalonia* as prey.

The difference between specialists and generalists is well shown by the studies at Chimney Rock. One nest of *basilaris* contained nine genera of prey belonging to seven families (including bees, wasps, an ant, and an ichneumon). In contrast, *inversus* has been reported as a specialist on male *Agapostemon* bees not only at Chimney Rock but in Alberta, Canada. Examples of prey specialization also include the Old World species *triangulum*, which has rarely been reported to use anything other than honey bees. But, as mentioned above, the bumblebee-wolf, *bicinctus*, proved capable of switching extensively to other bees and to *Podalonia* wasps during a period of bumble bee scarcity.

Some degree of plasticity in prey choice seems requisite in all species, since local conditions may frequently cause declines in certain prey species. Also, within its range, a species may be a member of several different communities with variably available prey. Considering prey

Table 10-3. Prey size niche overlap and taxonomic niche overlap for species-pair comparisons at Chimney Rock, Larimer Co., Colo.[1]

Prey size niche overlap				Taxonomic niche overlap				
	barb.	inv.	bas.	bic.		barb.	inv.	bas.	bic.
barbiger	----	----	----	----	barbiger	----	----	----	----
inversus	0.074	----	----	----	inversus	0.012	----	----	----
basilaris	0.303	0.861	----	----	basilaris	0.262	0.530	----	----
bicinctus	0.000	0.000	0.433	----	bicinctus	0.000	0.000	0.063	----

1. Combined data from all years of study. From O'Neill and Evans (1982).

selection from a broad perspective, O'Neill and Evans (1982) determined taxonomic niche breadths (on the generic level) for 30 populations of 16 species (based on sample sizes of 46–386 prey items). Populations showed a continuum from extreme specialists (a niche breadth of 1.00 in the Chimney Rock population of *bicinctus*) to broad generalists (a niche breadth of 12.39 in the Jackson Hole population of *zebratus* in 1967). Yearly variation was shown in the latter population, which had a niche breadth of only 4.31 in 1980. Different populations of one species sometimes occupied quite different positions on the specialist-generalist continuum. The best example is provided by *pulcher*; its Great Sand Dunes population had a niche breadth of 1.29, whereas its Jackson Hole population had niche breadths of 7.61, 9.42, and 11.95 in three different years.

In contrast, a tendency was noted for populations of certain species, though widely separated in time and space, to occupy similar positions on the specialist-generalist continuum. The population of *crabroniformis* we studied in Jackson Hole had a niche breadth of 1.51, whereas a population studied by Alcock (1974) in Seattle, Washington, had a niche breadth of 1.76. Widely spaced populations of *politus*, *solivagus*, and *albopilosus* also occupied similar positions on the continuum. The population of *inversus* we studied had a niche breadth of only 1.21, and it is clear that the population studied in Alberta by McCorquodale (personal communication, 1985) was similarly a specialist on male *Agapostemon* bees.

As Eurasian species of *Philanthus*, so far as known, restrict themselves to bees, it is of interest to note the extensive use of wasps by several North American species. In particular, *basilaris*, *zebratus*, *barbiger*, and *psyche* use many wasps (Table 10-4). On the other hand, wasps have never been recorded as prey of *bilunatus* or *lepidus* and only rarely as prey of *politus*, *crabroniformis*, or *multimaculatus*. Interpopulational differences have been reported, however; for example, *albopilosus* used many wasps at sites in Alberta and New York but few in Colorado and New Mexico (Table 7-8), and *pulcher* used many wasps in Jackson Hole, Wyoming, but rarely employed wasps at two Colorado sites (Tables 5-2 and 5-3).

Similarly, some species appear to prey exclusively or very nearly exclusively on males in preference to females (*inversus*, *albopilosus*) or on females in preference to males (*sanbornii*, *politus*). Interpopulational differences can also be cited in this regard, however; for example, Colorado populations of *multimaculatus* made extensive use of male bees, but females predominated in nests from Arizona (Table 4-7). Such

Table 10-4. Summary of type and sex of prey of Philanthus

Species	Total individuals	Total species	WASPS			BEES			Ratio ♂:♀
			% individuals	% species	% males	% individuals	% species	% males	
bicinctus	406	27	4	4	0	96	96	40	1:1.6
basilaris	221	54	31	46	25	69	54	56	1:1.2
zebratus	402	78	40	59	66	60	41	17	1:1.7
sanbornii	151	37	5	14	100	95	86	13	1:7
gibbosus	284	35	1	3	0	99	97	60	1.5:1
crabroniformis	529	32	1	13	57	99	87	62	1.7:1
barbatus	90	20	10	30	44	90	70	72	2.2:1
multimaculatus	151	12	2	8	0	98	92	71	2.3:1
inversus	168	3	0	0	0	100	100	100	168:0
pulcher	588	81	17	37	73	83	63	31	1:2.2
pacificus	128	27	26	48	35	74	52	70	1.7:1
barbiger	87	28	40	54	34	60	46	60	1:1
psyche	210	54	38	50	59	62	50	52	1.2:1
serrulatae	112	15	3	20	100	97	80	60	1.5:1
politus	379	36	1	3	100	99	97	22	1:3.5
ventilabris	48	20	4	15	50	96	85	39	1:1.5
lepidus	209	24	0	0	0	100	100	46	1:1.2
bilunatus	119	17	0	0	0	100	100	60	1.5:1
solivagus	250	41	7	17	0	93	83	41	1:1.7
albopilosus	216	33	19	45	85	81	55	94	11.7:1

differences may merely reflect local differences in prey availability rather than aspects of female prey choice.

Although we have characterized some species of *Philanthus* as prey specialists, the genus as a whole does not exhibit the degree of specialization characteristic of other Philanthinae. Many species of *Cerceris* prey on beetles of only one or a few related genera (Scullen and Wold, 1969); *Aphilanthops frigidus* takes only alate queen ants of the genus *Formica*; and species of *Clypeadon* and *Listropygia* take only worker ants of the genus *Pogonomyrmex* (Evans, 1962a). Alexander (1985) has shown that *Clypeadon laticinctus* females not only prey exclusively on workers of the western harvester ant, *Pogonmyrmex occidentalis*, but select workers of a particular size class. There is no more extreme example of prey specialization among the Sphecidae.

Intraspecific Variation in Female Reproductive Success

Chapter 9 summarized the evidence for intraspecific variation in male mating success. Corresponding evidence on females is available for some species of wasps, and again body size is the major known correlate of variation in nesting success and its components. For females of two species of wasps, one sphecid (Freeman, 1981) and one eumenid (Cowan, 1981), a positive correlation between body size and both lifetime reproductive success and size of young was discovered. Female digger wasps must go through a long sequence of tasks and solve a variety of problems (i.e., choose a nest site, construct a nest, provision the nest cells, produce and lay eggs, avoid predators, deal with the contingencies of the physical environment, and prevent parasitoids and cleptoparasites from destroying cell contents); thus overall variation in reproductive success could come from a variety of sources. There is evidence that variation in fitness may exist because larger females lay larger eggs (O'Neill, 1985), provision with larger prey (Byers, 1978; Linsley and MacSwain, 1956; O'Neill, 1985), provision at a higher rate (Alexander, 1985), or provision for longer periods each day because of thermal constraints on smaller females (Willmer, 1985). Data for the sphecids *Bembecinus quinquespinosus* (O'Neill and Evans, 1983b), *Sphecius grandis* (Hastings, 1986), and the beewolf *Philanthus pulcher* (Chapter 5) suggest that most females at the lower end of the size range may not provision nests at all. As with males, in order to determine the evolutionary significance of size advantages, we need to find out more about the genetic and environmental components of size variation. It is probable that selection pressures on female size are even stronger than those on male size, given the fact that female sphecids are usually larger on average than conspecific males.

The Role of Natural Enemies in the Evolution of Behavior

This chapter has repeatedly suggested that certain aspects of female behavior may have evolved in response to the attacks of cleptoparasites, parasitoids, and predators. That is by no means a novel conclusion or one applicable only to *Philanthus* (see, e.g., Spofford et al., 1986, and references therein). The following are some of the behaviors that we believe may have evolved in this context:

>   Nesting in dense aggregations
>   Leveling behavior

Nest closure
Construction of accessory burrows
Construction of short-term nests with few cells
Construction of a serrate profile in long-term nests
Prey carriage by middle legs beneath body
Devious approach flights
Removal of parasitized prey from nest

The Diptera and Hymenoptera that attack the nests of digger wasps are diverse in their behavior (Chapter 2), and we assume that the particular combination of antiparasite behaviors present in a given species reflects the past and present abundance of specific enemies. Nest closure, for example, may deter entry of hole searchers while delaying the wasp's entry long enough to permit a satellite fly to larviposit on the prey; diversionary approach flights may reduce the success of satellite flies but have no effect on hole searchers; and so forth.

In an interesting study of the cleptoparasitism of *Tachysphex terminatus* (Larrinae) by three species of miltogrammine flies, Spofford et al. (1986) showed that cell mortality varied from 30.6% in one population to 57.9% in a second population. This difference was largely a result of the differing abundance of the two species of *Senotainia* (*trilineata* and *vigilans*) at the two sites. *T. terminatus* has a variety of antiparasite behaviors, including nest closure, freeze stops, diversionary flights, and prey cleaning. These behaviors are less effective against *S. vigilans*, and the greater abundance of that species at the second site resulted in the greater cell mortality at that site. In particular, temporary closure of the nest, which occurred at both sites, facilitated larviposition by *S. vigilans*.

Species lacking well-defined antiparasite behavior often suffer high mortality. Two authors have noted about 50% mortality in the cells of the cicada killer *Sphecius speciosus* (Nyssoninae) (Evans, 1966c). In a population of *Podalonia occidentalis* (Sphecinae) fully 75% of the nest cells were parasitized by miltogrammine flies of four species (Evans, 1987).

Under each species of *Philanthus* we have presented whatever data we had on natural enemies, and it seems appropriate to summarize these data here (Table 10-5). Percentages of miltogrammine fly parasitism in several populations are arranged below in order from highest to lowest. Only samples of over 25 cells are included.

24 of 58 cells of *sanbornii* (Bedford, Mass.) 41%
11 of 37 cells of *gibbosus* (Rensselaerville, N.Y.) 30%

7 of 27 cells of *pulcher* (Larimer Co., Colo.) 26%
9 of 36 cells of *politus* (Rensselaerville, N.Y.) 25%
20 of 94 cells of *crabroniformis* (Jackson Hole, Wyo.) 21%
7 of 37 cells of *gibbosus* (Fort Collins, Colo.) 19%
5 of 31 cells of *bilunatus* (Larimer Co., Colo.) 16%
19 of 133 cells of *zebratus* (Jackson Hole, Wyo.) 14%
6 of 52 cells of *pulcher* (Jackson Hole, Wyo.) 12%
3 of 33 cells of *solivagus* (Rensselaerville, N.Y.) 9%
3 of 40 cells of *bicinctus* (Great Sand Dunes, Colo.) 8%

These percentages are no more than approximations based on the number of cells containing maggots, puparia, or fresh prey remains with no evidence of wasp larva or cocoon (fly puparia are not usually found in the cells, as the maggots dig into the soil to pupate). The percentages of parasitism show no obvious correlation to the presence or absence of described antiparasite behavior. However, flies of several

Table 10-5. Recorded nest parasitism in North American Philanthus[1]

Wasp species	Senotainia	Metopia	Phrosinella	Bombyliidae	Mutillidae	Chrysididae
bicinctus	O	O	-	-	O	O
basilaris	O	-	-	-	-	-
zebratus	●	O	●	-	-	-
sanbornii	●	●	O	O	●	-
gibbosus	●	●	O	●	●	O
crabroniformis	●	O	●	-	-	O
multimaculatus	O	-	-	-	-	-
inversus	O	-	-	-	-	-
pulcher	●	●	●	-	O	O
pacificus	O	-	-	-	-	-
barbiger	O	-	●	-	-	-
psyche	O	-	-	-	-	-
politus	●	-	●	-	-	-
ventilabris	-	●	-	-	-	-
lepidus	O	-	-	-	O	-
bilunatus	●	-	●	-	O	-
solivagus	O	-	O	-	-	-
albopilosus	-	-	●	-	-	-

1. O = seen in close association with nests or provisioning females.
 ● = reared from cells.

genera were involved, and there was no way to distinguish from cell contents whether the flies were satellites or hole searchers. It is interesting that *sanbornii*, which had the highest percentage of parasitism, neither closes the entrance nor uses known diversionary approach patterns. But the second species on the list, *gibbosus*, does indeed close the nest entrance while hunting and has a particularly prominent steplike and wavering approach flight. Different populations of some species show quite different percentages of parasitism. Some of the variation may represent sampling error, as the data come from short-term studies and parasite population levels may fluctuate. The generally high percentages indicate that nest parasites represent strong selection pressures on female beewolves, however.

Unfortunately there are few data on the success of Mutillidae as parasitoids of *Philanthus*. The one known example of parasitism by Bombyliidae (see under *gibbosus*, Chapter 4) suggests that parasitism by these flies is occasionally very high.

Problems for Further Study

As in our review of male mating strategies (Chapter 9), we conclude our discussion of female behavior with suggestions for further study. We have emphasized the importance of natural enemies as selective agents in the evolution of several aspects of nesting behavior; however, much remains to be done to demonstrate (or refute) the points we have made. Many more quantitative, observational data are needed on the behavior of the diverse Diptera and Hymenoptera that attack *Philanthus* and other digger wasps as well as on the responses of the wasps to these enemies. As of now, evasive approach flights have not been described for several beewolf species (Table 10-2), and more precise data are needed for all species. To what extent does the behavior of a wasp that is followed by a satellite fly differ from that of a wasp that is not followed? Does the presence of a closure truly increase the probability that a satellite fly will larviposit successfully? How do different percentages of parasitism in different populations correlate with observed behavioral differences on the part of the wasps? (Careful nest dissections and rearing of cell contents are needed to obtain data on amount and type of parasitism). Experimental manipulations might include, for example, alterations in nest density (by culling of females), filling of accessory burrows or making of additional ones, and making or removal of nest closures. The use of models or of dead flies manipulated by hand should also prove instructive.

Prey choice should be evaluated in relation to prey size and type, to niche overlap with coexisting species, and to temporal and spatial variation in prey availability. Variation in female nesting success should be searched for and examined with respect not only to parasite pressure but to size and age differences among females.

It is important that future studies include more precise measurements of nest density than have so far been made, as nest density is believed to be a critical factor not only in the incidence of natural enemies (and defenses against them) but also in influencing male mating systems. The factors that influence choice of nest sites should also be measured with appropriate instrumentation.

Finally, we hope that the behavior of species that are currently unstudied or incompletely studied will eventually be inventoried. Knowledge of more Eurasian and African species and members of other genera of Philanthinae should provide a sounder basis for understanding the behavior of *Philanthus* and may reveal previously unknown behaviors.

A Final Look at the Beewolves

Clearly, beewolves are among the most successful of solitary wasps. They are nearly worldwide in distribution, and the 34 North American species collectively range widely on the continent, from southern Canada well into Mexico and from sea level to at least 2800 m in elevation. Although all species require soil for nesting that is more or less bare and at least moderately friable, the variety of habitats they occupy is considerable: clearings in forests, eroded banks and hillsides, areas of alluvial or windblown sand, and various situations created by humans, such as dirt paths and roads, sand and gravel pits, and the margins of agricultural land. It is probable that beewolf populations have increased in recent years because of humankind's tendency to create areas of bare soil where none existed previously.

Two species, *gibbosus* and *ventilabris*, occur almost throughout the entire North American range of the genus; evidently they are able to thrive in diverse climates and soil types. At the other extreme, some species have quite limited ranges. Although *bicinctus* is an unusually large and showy wasp, for example, only four aggregations are known to us, all at elevations of 2000–2800 m in the Rocky Mountains, and all in coarse, sandy gravel with sparse sagebrush and other vegetation. One would expect greater variation in the life history patterns and behavior of widely distributed species such as *gibbosus* compared with those of limited distribution. Indeed we have noted, in *gibbosus*, variation in the number of generations per year, in the kinds of nesting substrates employed (including vertical banks), in nest depth and cell pattern, and in other features. It would take much more detailed studies, however, to determine the precise behavioral and physiological parameters that permit some species to range widely and require that others remain geographically and ecologically restricted.

Major Ethological Attributes of Philanthinae and of Philanthus

Compared with other Sphecidae, beewolves have a number of derived features that place them among the most advanced of solitary wasps. Not only do they attack formidable prey and nest successfully in spite of a host of natural enemies, but some species show evidence of nest sharing, which could represent an early stage in the evolution of communal nesting—a stage that members of the related genus *Cerceris* may have passed through (Evans and Hook, 1982a, 1982b; Hook, 1987). One advanced behavior not occurring in *Philanthus* or other Philanthinae is progressive provisioning of the cells. In *Bembix* and other higher Nyssoninae (and in some species of *Ammophila* in the Sphecinae) prey is brought into the nest over several days, as the larva grows, resulting in mother-larva contact, which is never observed in Philanthinae. There is no evidence that such contact has led to cooperative associations between adults of the same or sequential generations, however. In contrast, nest sharing by adult females, even when transitory and facultative, may provide a starting point in the evolution of stable multifemale family groups, which "are likely to have been a major setting for the origin and elaboration of social interactions in the wasps" (West-Eberhard, 1978, p. 833).

Another behavioral characteristic of Philanthinae, in this case unique to the subfamily, is territorial scent marking (Chapter 9). Clypeal hairbrushes are present in males of nearly all Philanthinae and occur elsewhere in the Sphecidae only in a few isolated genera (e.g., in the nyssonine genus *Hoplisoides*, where their function is unknown). Thus we believe that scent marking is an ancestral feature of the Philanthinae and that groups lacking it (e.g., *P. albopilosus* and apparently many species of *Cerceris*) have lost it secondarily. Within the Philanthinae, the hair brushes are best developed and scent marking most pronounced in *Philanthus* (Chapter 9).

The use of Hymenoptera as prey is probably also ancestral in the Philanthinae. Although most Cercerini use beetles, several Eurasian species of *Cerceris* prey on bees, suggesting that beewolf behavior may have been characteristic of the original stock of this now very large genus. *Pseudoscolia*, which Bohart and Menke (1976) place on the same phyletic line as the Cercerini, also uses bees. The switch from bees to ants in the Aphilanthopsini was accompanied by other specializations: in digging behavior, prey carriage (in two of the genera), and extreme narrowing of prey selection. One species of *Aphilanthops* has been reported to use bees, suggesting a switch from bees to ants within the tribe (Evans, 1977). Outside the Philanthinae, bees and wasps are used as prey only in species of *Palarus* (Larrinae) and some Australian *Bembix* (Nyssoninae), ants only by the genus *Tracheloides* (Crabroninae).

The subfamily Philanthinae is well characterized ethologically as well as morphologically. In addition to the features discussed above, we should mention the consistent presence of multicellular nests in the ground, the manner of stinging prey a single time on the thoracic venter (so far as studied), and the practice of laying the egg somewhat loosely, longitudinally along the ventral side of the last prey placed in the cell.

The three major tribes also appear well characterized ethologically (see Table 1-1 for content of tribes and distribution of genera). Derived features of Cercerini include vertical burrows cleared by pushing of soil upward with the pygidial plate, with a resulting rim of soil around the entrance; use of beetles as prey in most species; and frequent nest sharing and reutilization. Derived features of Aphilanthopsini include digging by spraying of soil from the nest entrance, use of ants as prey, and development of unique methods of prey carriage. Philanthini appear to have retained more of the hypothesized generalized traits of the subfamily; Bohart and Menke (1976) drew the same conclusions on the basis of an analysis of 36 structural characters. Philanthini have evolved territorial scent marking to a higher degree than members of other tribes, however. Other derived features present in some species include leveling behavior, accessory burrows, and a limited amount of nest sharing paralleling that in some species of *Cerceris*.

As the main thrust of our monograph has been a comparative behavioral study of the North American species of *Philanthus*, it seems appropriate to consider how our conclusions compare with decisions that have been made concerning the grouping of species on the basis of structure.

Taxonomic Treatments of North American Beewolves

Efforts have been made in the past to divide *Philanthus* into several genera or subgenera (Bohart and Menke, 1976). The following groups have been split off from time to time:

Oclocletes: a name proposed by Nathan Banks in 1913 for several large species having eyes strongly convergent toward the top of the head.

Anthophilus: a name proposed by A. G. Dahlbom in 1844 for several small species having reduced cuticular sculpturing.

Pseudanthophilus: a name proposed by W. H. Ashmead in 1899 for a single species, *ventilabris*, having the pronotal ridge transversely grooved.

Epiphilanthus: another name proposed by Ashmead for a single species, *solivagus*, which has fine, contiguous punctures over much of the body.

Each of these groups was erected primarily on the basis of a single, relatively superficial feature. By default, species without any of these features were left in *Philanthus* in the strict sense. Fortunately, these splinter groups have now been abandoned. The recent tendency has been to seek more natural groupings based on these and other morphological features but to grant these groupings no formal taxonomic status. Bohart and Grissell's (1975) *zebratus* group (Chapter 3) is equivalent to *Oclocletes*, their *ventilabris* and *solivagus* groups equivalent to *Pseudanthophilus* and *Epiphilanthus* respectively (covered in our Chapter 7 as "other species"). Their *pacificus* and *politus* groups are more or less equivalent to *Anthophilus* (our Chapters 5 and 6). Bohart and Grissell placed several additional species in their *gibbosus* group (Chapter 4) and *lepidus* group (included in Chapter 7).

Ferguson (1983) moved *ventilabris* into the *politus* group on the basis of the presence of metapleural laminae in that species. This same feature has been used to justify separation of the *politus* and *pacificus* groups. Metapleural laminae, minute projections behind the base of the hind wings, have no known function.

Even though we have, for convenience, followed Bohart and Grissell's (1975) species-groups in the text, in fact few concordances are apparent between behavioral features and this grouping. One exception is that the *pacificus* and *politus* groups, taken together (*Anthophilus*), do exhibit a unique nest type and are the only *Philanthus* in which mound leveling has been described (Table 10-1). Both Bohart and Grissell (1975) and Ferguson (1983) included *albopilosus* in this group. Nest type and the occurrence of partial mound leveling in this species suggest that it may, in fact, be a derivative of this group with special adaptations for its sand dune habitat (including the absence of territoriality and scent marking by males). The two species *lepidus* and *bilunatus* (Bohart and Grissell's *lepidus* group) have very similar female behavior (male behavior of *lepidus* is unstudied), but there seems nothing distinctive about their nesting behavior other than the consistent construction of accessory burrows.

Surprisingly, the *zebratus* group, though structurally different enough to have sometimes been granted full generic status (*Oclocletes*), seems to have no behavioral features clearly separating it from members of the *gibbosus* group. Members of the latter group appear to use more males than females as prey, whereas members of the *zebratus* group appear to use more females (Table 10-4); this analysis is based on more than 2400 prey records, but we doubt that it reflects a significant phylogenetic trend. Neither *ventilabris* nor *solivagus* seems individualistic enough in behavior to justify separation from a combined *zebratus-gibbosus-lepidus* group.

Behavior and the Evolution of Beewolves

Despite our conclusion that few concordances can be found between structure and behavior within the genus, and despite the greater lability of behavior compared with structure, we would be remiss if we did not conclude with a summary of what we believe to be trends in evolution within the genus and their apparent adaptive significance. Using comparisons with other digger wasps (especially other Philanthinae), we conceive of the ancestral beewolf as having the following characteristics:

1. Soil is scraped into an elongate mound that is not leveled (p. 234).
2. A temporary closure of the nest entrance is maintained (p. 235).
3. Accessory burrows are not constructed (p. 237).
4. Nests are proclinate with a diffuse cell pattern (p. 239).
5. Prey consists of a variety of bees and wasps (p. 246).
6. Territorial scent marking is a major component of male mating behavior (p. 214).

Not surprisingly, none of the species we have studied fits all of these criteria, although several differ in only one or two particulars. As pointed out in our introductory chapter, it seems probable that *Philanthus* had its origin in the Eastern Hemisphere. A fossil *Philanthus* has been described from the North American mid-Tertiary (Bohart and Menke, 1976), however; so it is clear that the genus has been present on this continent for at least about 25 million years. The genus has diversified to produce 34 species and to give rise to a neotropical derivative, the genus *Trachypus*. Adaptive radiation has resulted in species as large as *bicinctus* (females averaging 22 mm) and as small as *parkeri* (females averaging 7–8 mm). Corresponding differences in the size of the prey employed, as well as differences in emergence time, have enhanced the ability of some species to share the same habitat. It is probable that, in general, bare soil of suitable quality has been a more important limiting factor than prey availability, considering the diversity of prey used by many species and the ability of even apparent specialists to switch to other prey when the primary prey becomes reduced in numbers (e.g., *bicinctus*, Chapter 3).

A major aspect of adaptive radiation in *Philanthus* has been the development of a diversity of female behavior patterns functioning to reduce the success of several kinds of non-host-specific natural enemies (see p. 248 and Tables 10-1, 10-2, and 10-5). That different species have evolved different combinations of these behaviors we believe to be the result of the particular array of natural enemies most prevalent in the area in which they evolved. Even today, nest closure, construction of accessory burrows, and performance of diversionary flights are subject to considerable intraspecific variation, probably reflecting, in part at least, the relative local abundance of specific parasites and predators.

Several of the behaviors that, in various ways, function to reduce the success of natural enemies have evolved independently in unrelated groups of Sphecidae. Mound leveling in species such as *Philanthus pulcher* is strikingly similar to that occurring in *Bembix* and other higher Nyssoninae. Accessory burrows have been reported in species of Sphecinae and in many higher Nyssoninae, though perhaps more consistently present in *Philanthus* than elsewhere. Diversionary approaches to the nest have been reported in species of Astatinae, Larrinae, Crabroninae, and Nyssoninae; though varying in form, some are very similar to those occurring in *Philanthus*. Some of the same species of miltogrammine flies attack Sphecidae of all subfamilies, as well as wasps of other families. Wasps of groups that are diverse morphologically and in other aspects of behavior have thus converged on a set of behavioral modifications that have proved effective in permitting greater survival of their offspring.

Whereas females invest most of their time and energy in the production of offspring, males allocate most of their energies to seeking mates (Chapter 9). In *Philanthus*, where nesting aggregations tend to persist from year to year, it is to a male's advantage to remain at or near the site where he emerged (rather than, for example, patrolling the widely dispersed flowers where females seek prey and both sexes seek nectar). The distribution of territories is correlated with nest density, which varies between species and, to a degree, between populations of one species (pp. 205–210). Differences between species in rates of scent marking (Table 9-4) and in types of aggressive interactions with conspecific males have been described. Differences in pheromone chemistry have been demonstrated in the case of *bicinctus* and *basilaris* and will probably be found to occur generally.

Variability in mating tactics occurs in several species and is largely related to size. Smaller males, unable to compete successfully for the most desirable territories, may achieve copulations by intercepting females elsewhere or by taking over unoccupied or marginal territories. The most striking example of alternative tactics occurs in the dense population of *zebratus* at Deadman's Bar, in Grand Teton National Park, Wyoming. In this population larger males patrol above the nesting area (a nonaggressive tactic) while smaller males defend territories on the periphery. That some males were seen to undertake both tactics indicates that the alternative behaviors are not the result of a genetic polymorphism. In another site some 40 km away, where the population was smaller and nests more scattered, we noted no patrolling males; but several males had established territories outside the nesting area.

Finally, in one species, *albopilosus*, we found no evidence of male territoriality or scent marking. This species lacks clypeal brushes and has reduced mandibular glands. We believe that these structures have been secondarily reduced and that territoriality has been replaced by a form of patrolling more adaptive for the sand dune habitat of this species, where there are few if any plants that might be scent-marked.

Although we have found it convenient to discuss male behavior and female behavior separately, clearly they are interrelated as well as affected by similar intrinsic and extrinsic factors. In Figure 10-3 we have diagrammed some of the factors we believe to have been important in influencing the evolution of these wasps.

Thriving aggregations of beewolves present scenes of high drama,

Major Features of Nesting Behavior 259

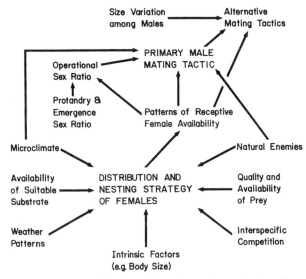

Figure 10-3. A flow chart, based on research on beewolves and other digger wasps, that outlines biotic and abiotic factors that may strongly influence the evolution of male and female behavior. Although significance of some of the factors (e.g., interspecific competition) is largely unsubstantiated, the importance of others (e.g., natural enemies, size variation) seems assured.

with constant comings and goings and diverse interactions between conspecifics and between *Philanthus* and prey, parasites, and intruders. The problem is to understand the script and, beyond that, to determine how it came to be written. For persons like us, who like to spend their summers in the field watching animals and asking why they act as they do, beewolves remain a continuing challenge.

References

Alcock, J. 1974. The behaviour of *Philanthus crabroniformis* (Hymenoptera: Sphecidae). *J. Zool. (London)* 173: 233–246.

Alcock, J. 1975a. The nesting behavior of some sphecid wasps of Arizona, including *Bembix, Microbembex,* and *Philanthus. J. Ariz. Acad. Sci.* 10: 160–165.

Alcock, J. 1975b. The nesting behavior of *Philanthus multimaculatus* Cameron (Hymenoptera, Sphecidae). *Am. Midl. Nat.* 93: 222–226.

Alcock, J. 1975c. Male mating strategies of some philanthine wasps (Hymenoptera: Sphecidae). *J. Kans. Entomol. Soc.* 48: 532–545.

Alcock, J. 1975d. Territorial behavior by males of *Philanthus multimaculatus* (Hymenoptera: Sphecidae) with a review of territoriality in male sphecids. *Anim. Behav.* 23: 889–895.

Alcock, J. 1975e. The behaviour of western cicada killer males, *Sphecius grandis* (Sphecidae, Hymenoptera). *J. Nat. Hist.* 9: 561–566.

Alcock, J. 1978. Notes on male mate-locating behavior in some bees and wasps of Arizona. *Pan-Pac. Entomol.* 54: 215–225.

Alcock, J. 1979. The relation between female body size and provisioning behavior in the bee *Centris pallida* Fox (Hymenoptera: Anthophoridae). *J. Kans. Entomol. Soc.* 52: 623–632.

Alcock, J. 1984. Long-term maintenance of size variation in populations of *Centris pallida* (Hymenoptera: Anthophoridae). *Evolution* 38: 220–223.

Alcock, J., and G. J. Gamboa. 1975. Home ranges of male *Cerceris simplex macrosticta* (Hymenoptera, Sphecidae). *Psyche (Camb., Mass.)* 81: 528–533.

Alcock, J., E.M. Barrows, G. Gordh, L.J. Hubbard, L. Kirkendall, D.W. Pyle, T.L. Ponder, and F.G. Zalom. 1978. The ecology and evolution of male reproductive behaviour in the bees and wasps. *Zool. J. Linn. Soc.* 64: 293–326.

Alexander, B. 1985. Predator-prey interactions between the digger wasp *Clypeadon laticinctus* and the harvester ant *Pogonomyrmex occidentalis. J. Nat. Hist.* 19: 1139–1154.

Alexander, B. 1986. Alternative methods of nest provisioning in the digger wasp *Clypeadon laticinctus* (Hymenoptera: Sphecidae). *J. Kans. Entomol. Soc.* 59: 59–63.

Armitage, K.B. 1965. Notes on the biology of *Philanthus bicinctus* (Hymenoptera: Sphecidae). *J. Kans. Entomol. Soc.* 38: 89–100.

Austad, S.N. 1984. A classification of alternative reproductive behaviors and methods for field-testing ESS models. *Am. Zool.* 24: 309–320.

Austad, S.N., and R.D. Howard, eds. 1984. Alternative reproductive tactics. *Am. Zool.* 24: 307–418.

Barrows, E.M., and T.P. Snyder. 1973. Halictine bee prey of *Philanthus gibbosus* (Hymenoptera: Sphecidae) in Kansas. *Entomol. News* 84:314–316.

Bateson, P., ed. 1983. *Mate Choice*. Cambridge, England: Cambridge Univ. Press. 462 pp.

Berland, L. 1925. Hyménoptères Vespiformes I. Vol. 10 of *Faune de France*. Paris: Lechevalier. 364 pp.

Betts, A.D. 1936. The bee pirate (*Philanthus triangulum* F.). *Bee World* 17: 64–66.

Beusekom, G.J. van. 1948. Some experiments on the optical orientation in *Philanthus triangulum* Fabr. *Behaviour* 1: 195–225.

Boake, C.R.B. 1986. A method for testing adaptive hypotheses of mate choice. *Am. Nat.* 127: 654–666.

Bohart, G.E. 1954. Honey bees attacked at their hive entrance by the wasp *Philanthus flavifrons* Cresson. *Proc. Entomol. Soc. Wash.* 56: 26–27.

Bohart, G.E., and G.F. Knowlton. 1953. Notes on mating, prey provisioning and nesting in *Sphex procerus* (Dahlbom). *Proc. Entomol. Soc. Wash.* 55: 100–101.

Bohart, R.M., and E.E. Grissell. 1975. California wasps of the subfamily Philanthinae (Hymenoptera: Sphecidae). *Bull. Calif. Insect. Surv.* 19:1–92.

Bohart, R.M., and L.S. Kimsey. 1982. A synopsis of the Chrysididae in America north of Mexico. *Mem. Am. Entomol. Soc.*, no. 33. 266 pp.

Bohart, R.M., and A.S. Menke. 1976. *Sphecid Wasps of the World: A Generic Revision*. Berkeley: Univ. of California Press. 695 pp.

Borg-Karlson, A.K., and J. Tengö. 1980. Pyrazines as marking volatiles in philanthine and nyssonine wasps (Hymenoptera: Sphecidae). *J. Chem. Ecol.* 6: 827–835.

Bouvier, E.L. 1900. Les variations des habitudes chez les Philanthes. *C. R. Soc. Biol. Paris* 52: 1129–1131.

Bouvier, E.L. 1916. Quelques observations sur les Philanthes. *Ann. Inst. Pasteur (Paris)* 30: 205–208.

Bradbury, J.W. 1981. The evolution of leks. In R.D. Alexander and D. W. Tinkle, eds., *Natural Selection and Social Behavior: Recent Research and Theory*, pp. 138–169. New York: Chiron Press.

Brockmann, H.J. 1979. Nest-site selection in the great golden digger wasp, *Sphex ichneumoneus* L. (Sphecidae). *Ecol. Entomol.* 4: 211–224.

Brothers, D.J. 1975. Phylogeny and classification of the aculeate Hymenoptera, with special reference to Mutillidae. *Univ. Kans. Sci. Bull.* 50: 483–648.

Brown, J.L. 1964. The evolution of diversity in avian territorial systems. *Wilson Bull.* 6: 160–169.

Brown, J.L. 1975. *The Evolution of Behavior*. New York: W.W. Norton. 761 pp.

Brückner, D. 1978. Why are there inbreeding effects in haplo-diploid systems? *Evolution* 32: 456–458.

Byers, G.W. 1978. Nests, prey, behavior, and development of *Cerceris halone* (Hymenoptera: Sphecidae). *J. Kans. Entomol. Soc.* 51: 818–831.

Campbell, G.S. 1977. *An Introduction to Environmental Biophysics*. New York: Springer-Verlag. 159 pp.

Cazier, M.A., and M.A. Mortenson. 1965. Studies on the bionomics of sphecoid wasps. II. *Philanthus gibbosus* (Fabricius) and *Philanthus anna* Dunning (Hymenoptera: Sphecidae). *Bull. South. Calif. Acad. Sci.* 64: 171–206.

Coville, R.E., and P.L. Coville. 1980. Nesting biology and male behavior of *Trypoxylon (Trypargilum) tenoctitlan* in Costa Rica (Hymenoptera: Sphecidae). *Ann. Entomol. Soc. Am.* 73: 110–119.

Coville, R.E., and C. Griswold. 1984. Biology of *Trypoxylon (Trypargilum) superbum* (Hymenoptera: Sphecidae), a spider-hunting wasp with extended guarding of the brood by males. *J. Kans. Entomol. Soc.* 57: 365–376.

Cowan, D.P. 1981. Parental investment in two solitary wasps, *Ancistrocerus adiabatus* and *Euodynerus foraminatus* (Eumenidae: Hymenoptera). *Behav. Ecol. Sociobiol.* 9: 95–102.

Cross, E.A., M.G. Stith, and T.R. Bauman. 1975. Bionomics of the organ-pipe muddauber, *Trypoxylon politum* (Hymenoptera: Sphecoidea). *Ann. Entomol. Soc. Am.* 68: 901–916.

Cross, E.A., A.E.-S. Mostafa, T.R. Bauman, and I.J. Lancaster. 1978. Some aspects of energy transfer between the organ pipe mud-dauber *Trypoxylon politum* and its araneid spider prey. *Environ. Entomol.* 7: 647–652.

Darwin, C. 1859. *On the Origin of Species*. Facsimile of the first edition. Cambridge, Mass.: Harvard University Press, 1964. 502 pp.

Dawkins, R. 1980. Good strategy or evolutionarily stable strategy? In G.W. Barlow and J. Silverberg, eds., *Sociobiology: Beyond Nature/Nurture?*, pp. 331–367. Boulder, Colo.: Westview Press.

Dominey, W.J. 1984. Alternative mating tactics and evolutionarily stable strategies. *Am. Zool.* 24: 385–396.

Duffield, R.M., M. Shamim, J.W. Wheeler, and A.S. Menke. 1981. Alkylpyrazines in the mandibular gland secretions of *Ammophila* wasps (Hymenoptera: Sphecidae). *Comp. Biochem. Physiol. B Comp. Biochem.* 70: 317–318.

Dunbar, R.I.M. 1982. Intraspecific variation in mating strategy. In P.P.G. Bateson and P.H. Klopfer, eds., *Perspectives in Ethology*, vol. 5, pp. 385–431. New York: Plenum Press.

Dunning, S.N. 1898. Notes on *Philanthus*. *Can. Entomol.* 30: 152–155.

Dutt, G.R. 1914. Life history of Indian insects (Hymenoptera). *Mem. Dept. Agric. India, Entomol. Ser., Agric. Res. Inst., Pusa* 4: 183–267.

Eberhard, W.G. 1974. The natural history and behaviour of the wasp *Trigonopsis cameronii* Kohl (Sphecidae). *Trans. R. Entomol. Soc. Lond.* 125: 295–328.

Edmunds, M. 1974. *Defence in Animals*. Harlow, Essex, England: Longman. 357 pp.

Eickwort, G.C., and H.S. Ginsberg. 1980. Foraging and mating behavior in Apoidea. *Annu. Rev. Entomol.* 25: 421–446.

Elliott, N.B. 1984. Behavior of males of *Cerceris watlingensis* (Hymenoptera: Sphecidae, Philanthinae). *Am. Midl. Nat.* 112: 85–90.

Emlen, S.T. 1976. Lek organization and mating strategies in the bullfrog. *Behav. Ecol. Sociobiol.* 1: 283–313.

Emlen, S.T., and L.W. Oring. 1977. Ecology, sexual selection, and the evolution of mating systems. *Science (Wash., D.C.)* 197: 215–223.

Endo, A. 1980. The behavior of a miltogrammine fly *Metopia sauteri* (Townsend) (Diptera: Sarcophagidae) cleptoparasitizing on a spider wasp *Episyron arrogans* (Smith) (Hymenoptera: Pompilidae). *Kontyu* 48: 445–457.

Evans, H.E. 1955. *Philanthus sanbornii* as a predator of honeybees. *Bull. Brooklyn Entomol. Soc.* 50: 47.

Evans, H.E. 1957a. Studies on the larvae of digger wasps (Hymenoptera, Sphecidae). Part III: Philanthinae, Trypoxyloninae, and Crabroninae. *Trans. Am. Entomol. Soc.* 83: 79–117.

Evans, H.E. 1957b. *Studies on the Comparative Ethology of Digger Wasps of the Genus* Bembix. Ithaca, N.Y.: Cornell Univ. Press. 248 pp.

References

Evans, H.E. 1959. Studies on the larvae of digger wasps (Hymenoptera, Sphecidae). Part V: Conclusion. *Trans. Am. Entomol. Soc.* 85:137–191.

Evans, H.E. 1962a. A review of the nesting behaviour of digger wasps of the genus *Aphilanthops*, with special attention to the mechanics of prey carriage. *Behaviour* 19: 239–260.

Evans, H.E. 1962b. The evolution of prey-carrying mechanisms in wasps. *Evolution* 16: 468–483.

Evans, H.E. 1963. *Wasp Farm.* Garden City, N.Y.: Natural History Press, 178 pp.

Evans, H.E. 1964a. The classification and evolution of digger wasps as suggested by larval characters (Hymenoptera: Sphecoidea). *Entomol. News* 75: 225–237.

Evans, H.E. 1964b. Notes on the nesting behavior of *Philanthus lepidus* Cresson (Hymenoptera, Sphecidae). *Psyche (Camb., Mass.)* 71: 142–149.

Evans, H.E. 1964c. Further studies on the larvae of digger wasps (Hymenoptera: Sphecidae). *Trans. Am. Entomol. Soc.* 90: 235–299.

Evans, H.E. 1965. Simultaneous care of more than one nest by *Ammophila azteca* Cameron (Hymenoptera, Sphecidae). *Psyche (Camb., Mass.)* 72: 8–23.

Evans, H.E. 1966a. The accessory burrows of digger wasps. *Science (Wash. D.C.)* 152: 465–471.

Evans, H.E. 1966b. Nests and prey of two species of *Philanthus* in Jackson Hole, Wyoming (Hymenoptera, Sphecidae). *Great Basin Nat.* 26: 35–40.

Evans, H.E. 1966c. *The Comparative Ethology and Evolution of the Sand Wasps.* Cambridge, Mass.: Harvard Univ. Press. 526 pp.

Evans, H.E. 1970. Ecological-behavioral studies of the wasps of Jackson Hole, Wyoming. *Bull. Mus. Comp. Zool. Harv. Univ.* 140: 451–511.

Evans, H.E. 1973. Burrow sharing and nest transfer in the digger wasp *Philanthus gibbosus* (Fabricius). *Anim. Behav.* 21: 302–308.

Evans, H.E. 1974. Digger wasps as colonizers of new habitat (Hymenoptera: Aculeata). *J.N.Y. Entomol. Soc.* 82: 259–267.

Evans, H.E. 1975. Nesting behavior of *Philanthus albopilosus* with comparisons between two widely separated populations. *Ann. Entomol. Soc. Am.* 68: 888–892.

Evans, H.E. 1977. *Aphilanthops hispidus* as a predator on bees (Hymenoptera: Sphecidae). *Pan-Pac. Entomol.* 53: 123.

Evans, H.E. 1982. Nesting and territorial behavior of *Philanthus barbatus* Smith (Hymenoptera: Sphecidae). *J. Kans. Entomol. Soc.* 55: 571–576.

Evans, H.E. 1987. Observations on the prey and nests of *Podalonia occidentalis* Murray (Hymenoptera, Sphecidae). *Pan-Pac. Entomol.* 63: 130–134.

Evans, H.E., and A.W. Hook. 1982a. Communal nesting in Australian *Cerceris* digger wasps. In M.D. Breed, C.D. Michener, and H.E. Evans, eds., *The Biology of Social Insects*, pp. 159–163. Boulder, Colo.: Westview Press.

Evans, H.E., and A.W. Hook. 1982b. Communal nesting in the digger wasp *Cerceris australis* (Hymenoptera: Sphecidae). *Austr. J. Zool.* 30: 557–568.

Evans, H.E., and C.S. Lin. 1959. Biological observations on digger wasps of the genus *Philanthus* (Hymenoptera: Sphecidae). *Wasmann J. Biol.* 17: 115–132.

Evans, H.E., and R.W. Matthews. 1973a. Observations on the nesting behavior of *Trachypus petiolatus* (Spinola) in Colombia and Argentina (Hymenoptera: Sphecidae: Philanthini). *J. Kans. Entomol. Soc.* 46: 165–175.

Evans, H.E., and R.W. Matthews. 1973b. Systematics and nesting behavior of Australian *Bembix* sand wasps (Hymenoptera, Sphecidae). *Mem. Am. Entomol. Inst. (Gainesville)*, no. 20. 387 pp.

Evans, H.E., and K.M. O'Neill. 1978. Alternative mating strategies in the digger wasp *Philanthus zebratus* Cresson. *Proc. Natl. Acad. Sci. USA* 75: 1901–1903.

Evans, H.E., and K.M. O'Neill. 1985. Male territorial behavior in four species of the tribe Cercerini (Sphecidae: Philanthinae). *J. N.Y. Entomol. Soc.* 93: 1033–1040.

Evans, H.E., and K.M. O'Neill. 1986. The reproductive and nesting behavior of *Bembecinus nanus strenuus* (Mickel) (Hymenoptera, Sphecidae). *Proc. Entomol. Soc. Wash.* 88: 628–633.

Evans, H.E., and M.J. West-Eberhard. 1970. *The Wasps*. Ann Arbor: Univ. of Michigan Press. 265 pp.

Evans, H.E., K.M. O'Neill, and R.P. O'Neill. 1986. Nesting site changes and nocturnal clustering in the sand wasp *Bembecinus quinquespinosus* (Hymenoptera: Sphecidae). *J. Kans. Entomol. Soc.* 59: 280–286.

Fabre, J.H. 1891. *Souvenirs entomologiques*. Quatrième séries. Paris: Delagrave. 433 pp.

Fabricius, J.C. 1790. Nova insectorum genera. *Skr. Naturhist. Selskabet (Copenhagen)* 1: 213–228.

Fahringer, J. 1922. Die Feinde der Schlammfliege. *Z. Wiss. Insekt. Biol.* 17: 113–124.

Ferguson, G.R. 1983. Two new species of the genus *Philanthus* and a key to the *politus* group (Hymenoptera: Philanthidae). *Pan-Pac. Entomol.* 59: 55–63.

Ferguson, G.R. 1984. Revision of the *Philanthus zebratus* group (Hymenoptera: Philanthidae). *J. N.Y. Entomol. Soc.* 91: 289–303.

Ferton, C. 1905. Notes sur l'instinct des Hyménoptères mellifères et ravisseurs. Troisième séries. *Ann. Soc. Entomol. Fr.* 74: 56–104.

Freeman, B.E. 1980. A population study in Jamaica on adult *Sceliphron assimile* (Dahlbom) (Hymenoptera: Sphecidae). *Ecol. Entomol.* 5: 19–30.

Freeman, B.E. 1981. Parental investment, maternal size, and population dynamics of a solitary wasp. *Am. Nat.* 117: 357–362.

Gambino, P. 1985. A new record for *Philanthus neomexicanus* Strandtmann (Hymenoptera: Philanthidae) and some insects found in its burrow. *Pan-Pac. Entomol.* 61: 226.

Grandi, G. 1961. Studi di un entomologo sugli imenotteri superiori. *Boll. Ist. Entomol. Univ. Studi Bologna* 25: 1–659.

Gwynne, D.T. 1978. Male territoriality in the bumblebee wolf, *Philanthus bicinctus* (Mickel) (Hymenoptera, Sphecidae): Observations on the behaviour of individual males. *Z. Tierpsychol.* 47: 89–103.

Gwynne, D.T. 1980. Female defense polygyny in the bumblebee wolf, *Philanthus bicinctus* (Hymenoptera: Sphecidae). *Behav. Ecol. Sociobiol.* 7: 213–225.

Gwynne, D.T. 1981. Nesting biology of the bumblebee wolf *Philanthus bicinctus* Mickel (Hymenoptera: Sphecidae). *Am. Midl. Nat.* 105: 130–138.

Gwynne, D.T., and K.M. O'Neill. 1980. Territoriality in digger wasps results in sex biased predation on males (Hymenoptera: Sphecidae: *Philanthus*). *J. Kans. Entomol. Soc.* 53: 220–224.

Hager, B.J., and F.E. Kurczewski. 1985. Reproductive behavior of male *Ammophila harti* (Fernald) (Hymenoptera: Sphecidae). *Proc. Entomol. Soc. Wash.* 87: 597–605.

Halliday, T.R. 1983. The study of mate choice. In P. Bateson, ed., *Mate Choice*, pp. 3–32. Cambridge, England: Cambridge Univ. Press.

Hamm, A.H., and O.W. Richards. 1930. The biology of the British fossorial wasps of the families Mellinidae, Gorytidae, Philanthidae, Oxybelidae, and Trypoxylonidae. *Trans. Entomol. Soc. Lond.* 78: 95–131.

Hartman, C.G. 1944. A note on the habits of *Trypoxylon politum* Say (Hymenoptera: Sphecidae). *Entomol. News* 55: 7–8.

Hastings, J. 1986. Provisioning by female western cicada killer wasps, *Sphecius grandis* (Hymenoptera: Sphecidae): influence of body size and emergence time on individual provisioning success. *J. Kans. Entomol. Soc.* 59: 262–268.

Hefetz, A., and S.W.T. Batra. 1980. Chemistry of the cephalic secretions of eumenid wasps. *Comp. Biochem. Physiol. B Comp. Biochem.* 65: 455–456.

Hilchie, G.J. 1982. Evolutionary aspects of geographical variation in color and prey in the beewolf species *Philanthus albopilosus* Cresson. *Quaest. Entomol.* 18: 91–126.

Hook, A.W. 1984. Notes on the nesting and mating behavior of *Trypoxylon (Trypargilum) spinosum* (Hymenoptera: Sphecidae). *J. Kans. Entomol. Soc.* 57: 534–535.

Hook, A.W. 1985. The comparative ethology of communal nesting in the Philanthinae (Hymenoptera: Sphecidae). Ph.D. dissertation, Colorado State Univ. 142 pp.

Hook, A.W. 1987. Nesting behavior of Texas *Cerceris* digger wasps with emphasis on nest reutilization and nest sharing (Hymenoptera: Sphecidae). *Sociobiology* 13: 93–118.

Hook, A.W., and R.W. Matthews. 1980. Nesting biology of *Oxybelus sericeus* with a discussion of nest guarding by male sphecid wasps (Hymenoptera). *Psyche (Camb., Mass.)* 87: 21–37.

Iwasa, Y., F.J. Odendaal, D.D. Murphy, P.R. Ehrlich, and A.E. Launer. 1983. Emergence patterns in male butterflies: A hypothesis and a test. *Theor. Popul. Biol.* 23: 363–379.

Iwata, K. 1976. *Evolution of Instinct: Comparative Ethology of Hymenoptera*. New Delhi, India: Amerind Publ. Co. 535 pp.

Jayasingh, D.B. 1980. A new hypothesis on cell provisioning in solitary wasps. *Biol. J. Linn. Soc.* 13: 167–170.

Jayasingh, D.B., and C.A. Taffe. 1983. The biology of the eumenid mud-wasp *Pachodynerus nasidens* in trapnests. *Ecol. Entomol.* 7: 283–289.

Kimsey, L.S. 1978. Nesting and male behavior in *Dynatus nigripes spinolae* (Lepeletier) (Hymenoptera: Sphecidae). *Pan-Pac. Entomol.* 54: 65–68.

Krombein, K.V. 1936. Biological notes on some solitary wasps (Hymenoptera: Sphecidae). *Entomol. News* 47: 93–99.

Krombein, K.V. 1952. Preliminary annotated list of the wasps of Westmoreland State Park, Virginia, with notes on the genus *Thaumatodryinus* (Hymenoptera: Aculeata). *Trans. Am. Entomol. Soc.* 78: 89–100.

Krombein, K.V. 1956. Miscellaneous prey records of solitary wasps. II. (Hymenoptera: Aculeata). *Bull. Brooklyn Entomol. Soc.* 51: 42–44.

Krombein, K.V. 1967. *Trap-nesting Wasps and Bees: Life Histories, Nests and Associates*. Washington, D.C.: Smithsonian Inst. Press. 570 pp.

Krombein, K.V. 1972. Miscellaneous prey records of solitary wasps. VI. Notes on some species from Greece (Hymenoptera: Aculeata). *Proc. Entomol. Soc. Wash.* 74: 383–385.

Krombein, K.V. 1979. Superfamily Sphecoidea. In K.V. Krombein, P.D. Hurd, Jr., D.R. Smith, and B.D. Burks, eds., *Catalog of Hymenoptera in America North of Mexico*, vol. 2, pp. 1573–1740. Washington, D.C.: Smithsonian Inst. Press.

Krombein, K.V. 1981. Biosystematic studies of Ceylonese wasps, VIII: A monograph of the Philanthidae (Hymenoptera: Sphecoidea). *Smithson. Contrib. Zool.*, no. 343. 75 pp.

Kurczewski, F.E. 1966. Comparative behavior of male digger wasps of the genus *Tachysphex* (Hymenoptera: Sphecidae, Larrinae). *J. Kans. Entomol. Soc.* 39: 436–453.

Kurczewski, F.E. 1982. An additional study on the nesting behaviors of species of *Miscophus* (Hymenoptera: Sphecidae). *Proc. Entomol. Soc. Wash.* 84: 67-80.

Kurczewski, F.E., and E.J. Kurczewski. 1984. Mating and nesting behavior of *Tachytes intermedius* (Viereck) (Hymenoptera: Sphecidae). *Proc. Entomol. Soc. Wash.* 86: 176–184.

Kurczewski, F.E., and R.C. Miller. 1983. Nesting behavior of *Philanthus sanbornii* in Florida (Hymenoptera: Sphecidae). *Fla. Entomol.* 66: 199–206.

Lane, R.S., J.R. Anderson, and E. Rogers. 1986. Nest provisioning and related activities of the sand wasp, *Bembix americana comata* (Hymenoptera: Sphecidae). *Pan-Pac. Entomol.* 62: 258–268.

Latreille, P.A., 1799. Mémoire sur un insecte qui nourrit ses petits d'abeilles domestiques. *Bull. Soc. Philom. Paris* 31: 49–50.

Lin, N. 1963. Territorial behavior in the cicada killer wasp, *Sphecius speciosus* (Drury) (Hymenoptera: Sphecidae). I. *Behaviour* 20: 115–133.

Lin, N. 1968. A note on the sleeping behavior of *Philanthus gibbosus* (Fabricius) (Hymenoptera: Sphecidae). *Proc. Entomol. Soc. Wash.* 70: 10–12.

Lin, N. 1978. Defended hunting territories and hunting behavior of females of *Philanthus gibbosus* (Hymenoptera: Sphecidae). *Proc. Entomol. Soc. Wash.* 80: 234–239.

Lin, N., and C.D. Michener. 1972. Evolution of sociality in insects. *Q. Rev. Biol.* 47: 131–159.

Linsley, E.G. 1962. Sleeping aggregations of aculeate Hymenoptera—II. *Ann. Entomol. Soc. Am.* 55: 148–164.

Linsley, E.G., and J.W. MacSwain. 1956. Some observations on the nesting habits and prey of *Cerceris californica* Cresson (Hymenoptera, Sphecidae). *Ann. Entomol. Soc. Am.* 49: 71–84.

Low, B.S. 1978. Environmental uncertainties and the parental strategies of marsupials and placentals. *Am. Nat.* 112: 197–213.

Lüps, P. 1973. Beobachtungen an *Bembecinus tridens* (Fabricius) (Hymenoptera, Sphecidae). *Mitt. Schweiz. Entomol. Ges.* 46: 131–139.

McClintock, J.B. 1986. On estimating energetic values of prey: Implications for optimal diet models. *Oecologia (Berl.)* 70: 161–162.

McCorquodale, D.B. 1986. Digger wasp (Hymenoptera: Sphecidae) provisioning flights as a defence against a nest parasite, *Senotainia trilineata* (Diptera: Sarcophagidae). *Can. J. Zool.* 64: 1620–1627.

McDaniel, C.A., R.W. Howard, K.M. O'Neill, and J.O. Schmidt. 1987. Chemistry of male mandibular gland secretions of *Philanthus basilaris* Cresson and *Philanthus bicinctus* (Mickel) (Hymenoptera: Sphecidae). *J. Chem. Ecol.* 13: 227–235.

Mackey, A.P. 1977. Growth and development of larval Chironomidae. *Oikos* 28: 270–275.

Maksimovic, M. 1958. Experimental research on the influence of temperature upon the development and the population dynamics of the gypsy moth (*Liparis dispar* L.). *Posebno Izdanje Biol. Inst. Srbije (Belgrade)* 3: 1–95.

Mally, C.W. 1909. Bee pirates. *Cape Good Hope Agric. J.*, no. 3. 9 pp.

Malyshev, S.I. 1968. *Genesis of the Hymenoptera and the Phases of Their Evolution.* Translated and edited by O.W. Richards and B. Uvarov. London: Methuen. 319 pp.

Mason, L.G. 1965. Prey selection by a non-specific predator. *Evolution* 19: 259–260.

Michener, C.D. 1986. Family-group names among bees. *J. Kans. Entomol. Soc.* 59: 219–234.

Miller, R.C., and F.E. Kurczewski. 1972. A review of nesting behavior in the genus

Entomognathus, with notes on *E. memorialis* Banks (Hymenoptera, Sphecidae). *Psyche (Camb., Mass.)* 79: 61–78.

Miller, R.C., and F.E. Kurczewski. 1973. Intraspecific interactions in aggregations of *Lindenius* (Hymenoptera: Sphecidae, Crabroninae). *Insectes Soc.* 20: 365–378.

Nielsen, E.T. 1933. Sur les habitudes des Hyménoptères aculéates solitaries. III. (Sphegidae). *Entomol. Medd.* 18: 259–348.

Olberg, G. 1953. Der Bienenfeind *Philanthus* (Bienenwolf). *Akad. Verlagsges. Geest Portig K.-G. (Leipzig)* 94: 1–88.

Olberg, G. 1959. *Das Verhalten der Solitären Wespen Mitteleuropas (Vespidae, Pompilidae, Sphecidae)*. Berlin: Deutscher Verlag Wissenchaften. 402 pp.

O'Neill, K.M. 1979. Territorial behavior in males of *Philanthus psyche* (Hymenoptera, Sphecidae). *Psyche (Camb., Mass.)* 86: 19–43.

O'Neill, K.M. 1981. Male mating strategies and body size in three species of beewolves (Hymenoptera: Sphecidae; *Philanthus*). Ph.D. dissertation, Colorado State Univ. 132 pp.

O'Neill, K.M. 1983a. The significance of body size in territorial interactions of male beewolves (Hymenoptera: Sphecidae, *Philanthus*). *Anim. Behav.* 31: 404–411.

O'Neill, K.M. 1983b. Territoriality, body size, and spacing in males of the beewolf *Philanthus basilaris* (Hymenoptera: Sphecidae). *Behaviour* 86: 295–321.

O'Neill, K.M. 1985. Egg size, prey size, and sexual size dimorphism in digger wasps (Hymenoptera: Sphecidae). *Can. J. Zool.* 63: 2187–2193.

O'Neill, K.M., and H.E. Evans. 1981. Predation on conspecific males by females of the beewolf *Philanthus basilaris* Cresson (Hymenoptera: Sphecidae). *J. Kans. Entomol. Soc.* 54: 553–556.

O'Neill, K.M., and H.E. Evans. 1982. Patterns of prey use in four sympatric species of *Philanthus* (Hymenoptera: Sphecidae) with a review of prey selection in the genus. *J. Nat. Hist.* 16: 791–801.

O'Neill, K.M., and H.E. Evans. 1983a. Body size and alternative mating tactics in the beewolf *Philanthus zebratus* (Hymenoptera; Sphecidae). *Biol. J. Linn. Soc.* 20: 175–184.

O'Neill, K.M., and H.E. Evans. 1983b. Alternative male mating tactics in *Bembecinus quinquespinosus* (Hymenoptera: Sphecidae): Correlations with size and color variation. *Behav. Ecol. Sociobiol.* 14: 39–46.

O'Neill, K.M., and R.P. O'Neill. 1987. Thermal stress and microhabitat selection in territorial males of the digger wasp *Philanthus psyche* (Hymenoptera: Sphecidae). *J. Therm. Biol.* (in press).

Packer, L. 1985. Two social halictine bees from southern Mexico with a note on two bee hunting philanthine wasps (Hymenoptera: Halictidae and Sphecidae). *Pan-Pac. Entomol.* 61: 291–298.

Paetzel, M.M. 1973. Behavior of the male *Trypoxylon rubrocinctum*. *Pan-Pac. Entomol.* 49: 26–30.

Peckham, D.J. 1977. Reduction of miltogrammine cleptoparasitism by male *Oxybelus subulatus* (Hymenoptera: Sphecidae). *Ann. Entomol. Soc. Am.* 70: 823–828.

Peckham, D.J., F.E. Kurczewski, and D.B. Peckham. 1973. Nesting behavior of Nearctic species of *Oxybelus* (Hymenoptera: Sphecidae). *Ann. Entomol. Soc. Am.* 66: 647–661.

Peckham, G.W., and E.G. Peckham. 1898. On the instincts and habits of the solitary wasps. *Wis. Geol. Nat. Hist. Surv. Bull.* 2: 1–245.

Peckham, G.W., and E.G. Peckham. 1905. *Wasps Social and Solitary*. Boston: Houghton Mifflin. 311 pp.

Piek, T., and W. Spanjer. 1986. Chemistry and pharmacology of solitary wasp venoms. In T. Piek, ed., *Venoms of the Hymenoptera: Biochemical, Pharmacological and Behavioural Aspects*, pp. 161–307. London: Academic Press.
Powell, J.A. 1967. Behavior of ground nesting wasps of the genus *Nitelopterus*, particularly *N. californicus* (Hymenoptera: Sphecidae). *J. Kans. Entomol. Soc.* 40: 331–346.
Powell, J.A., and J.A. Chemsak. 1959. Some biological observations on *Philanthus politus pacificus* Cresson. *J. Kans. Entomol. Soc.* 32: 115–120.
Rathmayer, W. 1962a. Das Paralysierungsproblem beim Bienenwolf *Philanthus triangulum* F. (Hym. Sphec.). *Z. Vgl. Physiol.* 45: 413–462.
Rathmayer, W. 1962b. Paralysis caused by the digger wasp *Philanthus*. *Nature (Lond.)* 196: 1148–1151.
Rathmayer, W. 1966. The effect of the poison of spider- and diggerwasps on their prey (Hymenoptera: Pompilidae, Sphecidae). *Mem. Inst. Butantan, Simp. Internac.* 33: 651–658.
Rau, P., and N. Rau. 1918. *Wasp Studies Afield*. Princeton, N.J.: Princeton Univ. Press. 372 pp.
Reinhard, E.G. 1924. The life history and habits of the solitary wasp, *Philanthus gibbosus*. *Annu. Rept. Smithson. Inst.* 1922, pp. 363–376.
Reinhard, E.G. 1929, *The Witchery of Wasps*. New York: Century Co. 291 pp.
Ristich, S.S. 1956. The host relationship of a miltogrammid fly *Senotainia trilineata* (VDW). *Ohio J. Sci.* 56: 271.
Rubink, W.L., and K.M. O'Neill. 1980. Observations on the nesting behavior of three species of *Plenoculus* Fox (Hymenoptera: Sphecidae). *Pan-Pac. Entomol.* 56: 187–196.
Rubio Espina, E. 1975. Revisión del género *Trachypus* Klug (Hymenoptera: Sphecidae). *Rev. Fac. Agron. Univ. Zulia Maracaibo* 3: 7–87.
Schall, J.J., and E.R. Pianka. 1980. Evolution of escape behavior diversity. *Am. Nat.* 115: 551–566.
Schmidt, J.O., K.M. O'Neill, H.M. Fales, C.A. McDaniel, and R.W. Howard. 1985. Volatiles from the mandibular glands of male beewolves (Hymenoptera: Sphecidae, *Philanthus*) and their possible roles. *J. Chem. Ecol.* 11: 895–901.
Schoener, T.W. 1974. Resource partitioning in ecological communities. *Science (Wash. D.C.)* 185: 27–39.
Schöne, H., and J. Tengö. 1981. Competition of males, courtship behaviour and chemical communication in the digger wasp *Bembix rostrata* (Hymenoptera, Sphecidae). *Behaviour* 77: 44–66.
Scriber, J.M., and R.C. Lederhouse. 1983. Temperature as a factor in the development and feeding ecology of tiger swallowtail caterpillars, *Papilio glaucus* (Lepidoptera). *Oikos* 40: 95–102.
Scullen, H.A., and J.L. Wold. 1969. Biology of wasps of the tribe Cercerini, with a list of the Coleoptera used as prey. *Ann. Entomol. Soc. Am.* 62: 209–214.
Shappirio, D.G. 1948. Observations on the biology of some mutillid wasps (Hym.: Mutillidae). With new distribution records. *Bull. Brooklyn Entomol. Soc.* 43: 157–159.
Shorey, H.H. 1977. The adaptiveness of pheromone communication. In D. White, ed., *Proceedings of the XV International Congress of Entomology*, pp. 294–307. College Park, Md.: Entomological Society of America.
Simonthomas, R.T. 1966. A method of breeding *Philanthus triangulum* F. (Sphecidae, Hymenoptera). *Entomol. Ber. (Amst.)* 26: 114–116.

Simonthomas, R.T., and E.P.R. Poorter. 1972. Notes on the behaviour of males of *Philanthus triangulum* (F.) (Hymenoptera, Sphecidae). *Tijdschr. Entomol.* 115: 141–152.

Simonthomas, R.T., and A.M.J. Simonthomas. 1972. Some observations on the behaviour of females of *Philanthus triangulum* (F.) (Hymenoptera, Sphecidae). *Tijdschr. Entomol.* 115: 123–139.

Simonthomas, R.T., and A.M.J. Simonthomas. 1977. *A Pest of the Bee Wolf in the Apiculture of the Dakhla Oasis, Egypt*. Amsterdam, Netherlands: Pharmacological Lab. Univ. Amsterdam, 23 pp.

Simonthomas, R.T., and A.M.J. Simonthomas. 1980. *Philanthus triangulum* and its recent eruption as a predator of honeybees in an Egyptian oasis. *Bee World* 61: 97–107.

Simonthomas, R.T., and R.L. Veenendaal. 1978. Observations on the behaviour underground of *Philanthus triangulum* (Fabricius) (Hymenoptera, Sphecidae). *Entomol. Ber. (Amst.)* 38: 3–8.

Sower, L.L., R.S. Kaae, and H.H. Shorey. 1973. Sex pheromones of Lepidoptera. XLI. Factors limiting potential distance of sex pheromone communication in *Trichoplusia ni*. *Ann. Entomol. Soc. Am.* 66: 1121–1122.

Spofford, M.G., F.E. Kurczewski, and D.J. Peckham. 1986. Cleptoparasitism of *Tachysphex terminatus* (Hymenoptera: Sphecidae) by three species of Miltogrammini (Diptera: Sarcophagidae). *Ann. Entomol. Soc. Am.* 79: 350–358.

Steiner, A.L. 1975. Description of the territorial behavior of *Podalonia valida* (Hymenoptera, Sphecidae) females in southeast Arizona, with remarks on digger wasp territorial behavior. *Quaest. Entomol.* 11: 113–127.

Steiner, A.L. 1978. Observations on spacing, aggressive and lekking behavior of digger wasp males of *Eucerceris flavocincta* (Hymenoptera: Sphecidae; Cercerini). *J. Kans. Entomol. Soc.* 51: 492–498.

Steiner, A.L. 1986. Stinging behavior of solitary wasps. In T. Piek, ed., *Venoms of the Hymenoptera: Biochemical, Pharmacological and Behavioural Aspects*, pp. 63–160. London: Academic Press.

Stubblefield, W. 1983. Research notes. *Sphecos* 7: 1–2.

Stubblefield, W. 1985. Research notes. *Sphecos* 10: 5.

Sutherland, W.J. 1985. Measures of sexual selection. In R. Dawkins and M. Ridley, eds., *Oxford Surveys in Evolutionary Biology*, vol. 2, pp. 90–101. Oxford, England: Oxford Univ. Press.

Thiem, H. 1932. Die Bienenwolf-Plage im Kaligebiet der Werra und ihre Bekämpfung. *Dtsch. Bienenzucht Theorie Praxis* 40: 173–186.

Thornhill, R., and J. Alcock. 1983. *The Evolution of Insect Mating Systems*. Cambridge, Mass.: Harvard Univ. Press. 547 pp.

Tinbergen, N. 1932. Über die Orientierung des Bienenwolfes (*Philanthus triangulum* Fabr.). *Z. Vgl. Physiol.* 16: 305–335.

Tinbergen, N. 1935. Über die Orientierung des Bienenwolfes. II. Die Bienenjagd. *Z. Vgl. Physiol.* 21: 699–716.

Tinbergen, N. 1951. *The Study of Instinct*. London: Oxford Univ. Press. 228 pp.

Tinbergen, N. 1958. *Curious Naturalists*. New York: Basic Books. 280 pp.

Tinbergen, N. 1972. *The Animal in Its World: Explorations of an Ethologist, 1932–1972*. Cambridge, Mass.: Harvard Univ. Press. 343 pp.

Tinbergen, N., and W. Kruyt. 1938. Über die Orientierung des Bienenwolfes. III. Die Bevorzugung bestimmter Wegmarken. *Z. Vgl. Physiol.* 25: 292–334.

Trivers, R.L. 1972. Parental investment and sexual selection. In B. Campbell, ed., *Sexual Selection and the Descent of Man*, pp. 136–179. Chicago: Aldine.

Tsuneki, K. 1943. On the habit of *Philanthus coronatus* Fabricius (Hymenoptera, Philanthidae). *Mushi* 15: 33–36. In Japanese.

Tsuneki, K. 1956. Ethological studies on *Bembix niponica* Smith, with emphasis on the psychobiological analysis of behaviour inside the nest (Hymenoptera: Sphecidae). I. Biological part. *Mem. Fac. Lib. Arts Fukui Univ. Ser. II Nat. Sci.* 6: 77–179.

Tsuneki, K. 1963. Comparative studies on the nesting biology of the genus *Sphex* (s.l.) in East Asia (Hymenoptera, Sphecidae). *Mem. Fac. Lib. Arts Fukui Univ. Ser. II Nat. Sci.* 13: 13–78.

Viereck, H.L., and T.D.A. Cockerell. 1904. The Philanthidae of New Mexico. II. *J. N.Y. Entomol. Soc.* 12: 129–146.

Vinson, S.B., H.J. Williams, G.W. Frankie, J.W. Wheeler, M.S. Blum, and R.E. Coville. 1982. Mandibular glands of male *Centris adani* (Hymenoptera: Anthophoridae). Their morphology, chemical constitutents, and function in scent marking and territorial behavior. *J. Chem. Ecol.* 8: 319–327.

Waser, S.K., and M.L. Waterhouse. 1983. The establishment and maintenance of sex biases. In S.K. Waser, ed., *Social Behavior of Female Vertebrates*, pp. 19–38. New York: Academic Press.

Wcislo, W.T. 1984. Gregarious nesting of a digger wasp as a "selfish herd" response to a parasitic fly (Hymenoptera: Sphecidae; Diptera: Sarcophagidae). *Behav. Ecol. Sociobiol.* 15: 157–160.

Wcislo, W.T. 1986. Host discrimination by a cleptoparasitic fly, *Metopia campestris* (Fallen) (Diptera: Sarcophagidae: Miltogramminae). *J. Kans. Entomol. Soc.* 59: 82–88.

Werner, F.G. 1960. A note on the prey and nesting site of *Cerceris truncata* Cameron (Hymenoptera: Sphecidae: Philanthinae). *Psyche (Camb., Mass.)* 67: 43–44.

West-Eberhard, M.J. 1978. Polygyny and the evolution of social behavior in wasps. *J. Kans. Entomol. Soc.* 51: 832–856.

Wheeler, W.M. 1919. The parasitic Aculeata: A study in evolution. *Proc. Am. Philos. Soc.* 58: 1–40.

Wiklund, C., and T. Fagerström. 1977. Why do males emerge before females? A hypothesis to explain the incidence of protandry in butterflies. *Oecologia (Berl.)* 31: 153–158.

Willmer, P.G. 1985. Size effects on the hygrothermal balance and foraging patterns of a sphecid wasp, *Cerceris arenaria*. *Ecol. Entomol.* 10: 469–479.

Willmer, P.G., and D.M. Unwin. 1981. Field analyses of insect heat budgets: Reflectance, size and heating rates. *Oecologia (Berl.)* 50: 250–255.

Wilson, E.O. 1970. Chemical communication within animal species. In E. Sondheimer and J.B. Simeone, eds., *Chemical Ecology*, pp. 133–155. New York: Academic Press.

Index

Abdominal hair brushes, 5, 9, 32
Accessory burrows, 19–21, 30, 235–237, 249, 256–257; of *Philanthus albopilosus,* 185, 186, 191; of *P. barbatus,* 103–104; of *P. barbiger,* 137; of *P. basilaris,* 53–54; of *P. bicinctus,* 39; of *P. bilunatus,* 176–177; of *P. crabroniformis,* 95, 231; of *P. gibbosus,* 82–83, 231; of *P. lepidus,* 171–172; of *P. serrulatae,* 156; of *P. ventilabris,* 166; of *P. zebratus,* 65
Adaptive radiation, 257
Agapostemon, specialization on, 22, 113, 244–246
Aggression: among females, 39, 72, 123, 232; among males, 16–17, 215–217, 221–222. *See also* Territoriality
Alcock, J.: on approach flights, 240; on mating behavior, 11, 201, 203; on *Philanthus crabroniformis,* 91–92, 96–100 243, 246; on *P. gibbosus,* 78–80, 86, 89, 243; on *P. multimaculatus,* 16–17, 105–111; on *P. ventilabris,* 165–169; on territoriality, 16–17, 212–220, 225
Alexander, B., 219, 220, 232, 247, 248
Alternative mating tactics, 18, 62, 223–226, 258–259
Ammophila, 15, 55, 68, 202, 206, 253
Anthophilus, 255
Antiparasite behavior, 237, 248–251, 257. *See also* Accessory burrows; Approach flights; Leveling behavior
Ants. *See* Formicidae

Aphilanthops, 3, 26, 234, 240, 254; *A. frigidus,* 182, 232, 247
Aphilanthopsini, 3, 254
Apis cerana indica, 197
Apis mellifera, 6, 22–25, 30–31; as prey of *Philanthus bicinctus,* 42; as prey of *P. crabroniformis,* 91, 97–99; as prey of *P. crotoniphilus,* 114; as prey of *P. sanbornii,* 91, 97–99; as prey of *P. triangulum,* 192, 195
Approach flights, 240–243, 249, 257
Armitage, K. B., 24, 33, 37–42
Asilidae, as predators, 26–27, 226; on *Philanthus basilaris,* 38, 42, 57; on *P. bicinctus,* 38, 42; on *P. gibbosus,* 90; on *P. psyche,* 150; on *P. pulcher,* 130
Astatinae, 4, 257
Austad, S. N., 223

Barrows, E. M., 78, 87, 89
Bee flies. *See* Bombyliidae
Beekeeping, 6, 30–31
Bembecinus nanus, 143, 144, 206
Bembecinus quinquespinosus, 201–204, 206, 221, 222, 224, 227
Bembix, 202, 232, 234, 242, 253, 257; Australian species, 2, 254; *B. americana,* 158; *B. niponica,* 227; *B. pruinosa,* 140, 183, 188, 206; *B. rostrata,* 201, 204, 206, 226
Bicyrtes, 206, 210, 234
Bohart, G. E., 91, 97–99

273

Bohart, R. M.: on Chrysididae, 28; on classification, 2–4, 78, 115, 214, 254–256; on life histories, 91, 112, 131, 134, 230
Bombus, as prey, 22, 31, 33, 41–43, 56–57
Bombyliidae, as parasites, 27–28, 77, 90, 237, 250
Borg-Karlson, A. K., 13, 15, 194, 213
Bouvier, E. L., 193, 194
Braconidae, as prey, 133, 139, 151
Brothers, D. J., 2
Bumble bees. See *Bombus*

Cazier, M. A.: on *Philanthus gibbosus*, 78, 79, 85, 86, 88–90, 243; on *P. multimaculatus*, 105–106, 108–111, 243
Centris, 13, 212, 220
Ceratochrysis, 28, 101, 130
Cercerini, 3, 231, 254
Cerceris, 3, 36, 201, 220, 240, 242, 247; mating behavior, 206, 209, 210, 214, 226, 253, 254; nesting behavior, 26, 234, 236, 239
Chrysididae: as natural enemies, 28, 44, 90, 101, 130, 195, 234, 250; as prey, 126, 133, 151
Cicindela, as predator, 191
Cleptoparasites, 28–30, 43–44, 200, 218, 248. See also Chrysididae; *Hilarella hilarella*; *Metopia*; *Phrosinella*; *Senotainia*
Closure of nest: final, 75, 129, 162; temporary, 19, 21, 30, 234–236, 249, 256, 257
Clypeadon, 3, 209, 214, 234, 240; *C. laticinctus*, 154, 219, 220, 232, 247
Clypeal brushes, 5, 9–10, 12–15, 37, 214, 253
Color, significance of, 227
Conopidae, as parasites, 27, 70, 77, 130, 164, 195
Copulation. See Mating; Mating behavior; Mating strategies; Mating success; Mating tactics
Costs associated with mating tactics, 226–227
Coville, R. E., 200, 208
Crabro, 232, 243
Crabroninae, 4, 257
Cuckoo wasps. See Chrysididae

Darwin, C., 198–199
Dasymutilla, 44, 77, 90, 174, 178
Dawkins, R., 223
Digging behavior, 19, 233–234
Dutt, G. R., 197
Dynatus nigripes, 200, 208

Economic importance, 30–31
Edmunds, M., 19, 237
Emlen, S. T., 199, 203
Epiphilanthus, 255
Eremiasphecium, 2–3
Escape tactics, 242. See also Approach flights
Eucerceris, 3, 26, 209, 214, 234, 240; *E. arenaria*, 13; *E. flavocincta*, 210, 211, 221
Evolution of behavior, 214–215, 237–239, 253–254, 256–259
Exoprosopa fascipennis albocollaris, 90

Fabre, J. H., 1, 25, 192–193
False burrows. See Accessory burrows
Ferguson, G. F.: on classification, 32, 45, 140, 164, 255, 256; on prey records, 58, 67
Ferton, C., 196–197
Flies: as prey, 24, 75, 138–139; as natural enemies, see Bombyliidae; Conopidae; Miltogramminae
Floaters, 17, 18, 223–226; in *Philanthus barbiger*, 136; in *P. basilaris*, 48; in *P. crabroniformis*, 94; in *P. tarsatus*, 154
Formicidae: as natural enemies, 130, 219; as prey, 55, 247
Fossils, 257
Freeman, B. E., 203, 248
Freeze-stops, 69, 99–101, 110, 114, 240–243

Gambino, P., 24, 138–139
Grandi, G., 10, 192, 196–197
Gwynne, D. T.: on nesting behavior, 31, 39–44, 245; on *Philanthus bicinctus*, 11, 26; on reproductive behavior, 12, 17, 18, 33–38, 202, 215

Habitat and life history, 6, 10–11, 231–233
Hamm, A. H., 192, 195
Hedychridium, 28, 90, 101, 130
Hedychrum intermedium, 28, 31, 195
Hilarella hilarella, 89, 101
Hilchie, G. J., 183–185, 188, 189, 191
Honey bees. See *Apis mellifera*
Hook, A. W., 159, 166, 168, 200, 201, 239, 253
Hoplisoides, 253
Hunting behavior. See Prey capture

Ichneumonidae, as prey, 55, 68, 126, 133, 139, 151, 190
Iwasa, Y., 202
Iwata, K., 2, 21, 199, 230, 235–237

Index   275

Jayasingh, D. B., 218

Krombein, K. V., 3, 131, 165, 175, 179; on *Philanthus basalis basalis*, 197; on *P. gibbosus*, 78, 89; on *P. venustus*, 197
Kurczewski, F. E., 25, 31, 70–75, 202, 236

Larrinae, 4, 239, 257
Latreille, P. A., 1
Leks, 16, 50–52, 102, 210
Leveling behavior, 19, 30, 234–235, 248, 256; of *Philanthus albopilosus*, 185; of *P. pacificus*, 132; of *P. politus*, 159–160; of *P. psyche*, 147–149; of *P. pulcher*, 123–124
Lin, N., 5, 78, 79, 81, 87, 198, 200
Lindenius, 200, 206, 221
Listropygia, 3, 240, 247
Locality studies. *See* Orientation

McCorquodale, D. B., 86, 89, 112–114, 242–243
McDaniel, C. A., 13, 15
Malaxation of prey, 25
Mandibular glands, 5, 9–10, 12–15, 37
Mason, L. G., 33, 42
Mating, frequency of, 11, 18, 200, 201
Mating behavior, 18, 204, 222–223; of *Philanthus albopilosus*, 184; of *P. basilaris*, 52–53; of *P. bicinctus*, 18, 37, 38; of *P. crabroniformis*, 92, 95; of *P. multimaculatus*, 106, 108; of *P. psyche*, 146; of *P. pulcher*, 117, 120–121; of *P. zebratus*, 61–63
Mating strategies, 198–229, 258. *See also* Floaters; Patrolling; Satellite males; Territoriality
Mating success, variation in, 220–223
Mating tactics, alternative, 18, 62, 223–226, 258–259
Megaselia, 182
Melittidae, as prey, 190
Methods of study, 6–7
Metopia, 30, 70, 195, 232, 237, 250; *M. argyrocephala*, 30, 43, 77, 89, 101, 130, 169
Microbembex, 56, 140, 154, 183, 189, 206, 234
Miltogramminae, 28–30, 43–44, 57, 257. *See also Hilarella hilarella; Metopia; Senotainia*
Mutillidae: as parasites, 28, 44–45, 77, 90, 174, 178, 196, 238, 250; as prey, 151, 157

Natural enemies, 26–30, 233, 248–251, 259. *See also* Asilidae; Bombyliidae; Chrysididae; Conopidae; Miltogramminae
Nest density, 204–211, 232, 248
Nest depth, variation in, 85, 149, 161, 173, 188, 238–239
Nesting behavior, 19–26, 230–251, 256–257; male participation in, 199–201; of *Philanthus albopilosus*, 184–191; of *P. barbatus*, 103–105; of *P. barbiger*, 137–139; of *P. basilaris*, 53–57; of *P. bicinctus*, 38–42; of *P. bilunatus*, 176–178; of *P. crabroniformis*, 95–100; of *P. gibbosus*, 82–89; of *P. inversus*, 112–113; of *P. lepidus*, 170–174; of *P. multimaculatus*, 108–111; of *P. pacificus*, 131–133; of *P. politus*, 159–163; of *P. psyche*, 147–150; of *P. pulcher*, 123–129; of *P. sanbornii*, 72–75; of *P. serrulatae*, 156–158; of *P. solivagus*, 179–182; of *P. triangulum*, 194–195; of *P. ventilabris*, 166–169; of *P. zebratus*, 65–67
Nest reutilization, 81–84, 195, 239
Nest sharing, 81–82, 168, 236, 239
Nest structure, 20–22, 235; evolution of, 237–239, 257. *See also* Nesting behavior
Niche breadth, 245–246
Nielsen, E. T., 21–22
Nyssoninae, 4–5, 15, 239, 253, 257

Oclocletes, 32, 255, 256
Olberg, G., 19, 26, 27, 192, 231, 234
Orientation, 19–20, 40, 65, 72, 85, 124
Oviposition, 21, 25–26
Oxybelus, 199–201, 206, 208, 214, 226

Packer, L., 89
Parasitoids, 27–28, 248. *See also* Bombyliidae; Conopidae; Mutillidae
Parker, F. D., 97, 231–232
Patrolling, 59–62, 205–207, 210, 224–225
Peckham, D. J., 200, 207, 208
Peckham, G. W.: on *Philanthus gibbosus*, 78–82, 86–89; on *P. ventilabris*, 165–168
Perilampidae, as prey, 139
Pheromones, 5, 9, 12–16, 37, 52, 211, 258
Philanthinae, 2–5, 231, 239, 253–254
Philanthinus, 2–3
Philanthus: distribution, 2–3, 252–253, 257; habitat and life history, 6, 10–11, 231–233; relationships, 4; structure and color, 8–10; taxonomy, 2–3, 255–256

276   Index

Philanthus albopilosus, 183–191; classification, 140, 256; comparative nesting behavior, 235, 246, 247, 250, 256; comparative reproductive behavior, 205–206, 214, 227, 231, 258
Philanthus barbatus, 101–105; comparative nesting behavior, 231, 235, 242, 247; comparative reproductive behavior, 207, 209, 210, 213, 214
Philanthus barbiger, 125, 134–139; comparative nesting behavior, 232, 235, 240, 242–247, 250; comparative reproductive behavior, 206, 209, 210, 213, 217, 227
Philanthus basalis basalis, 197
Philanthus basilaris, 45–57; comparative nesting behavior, 232, 235, 242, 243–247, 250; comparative reproductive behavior, 209–215, 217, 218, 221–226; predatory behavior, 24, 26, 38, 43; scent marking and territoriality, 13–16
Philanthus bicinctus, 8–9, 27, 32–45, 252, 257; comparative nesting behavior, 231, 233, 235, 242–247, 250; comparative reproductive behavior, 200, 201, 205, 208, 210–218, 221, 222, 226; predatory behavior, 22, 31; scent marking and territoriality, 13, 15, 17, 18
Philanthus bilunatus, 167, 175–178; comparative nesting behavior, 231, 235, 242, 246–247, 250, 256; comparative reproductive behavior, 207, 209, 210, 213, 227
Philanthus coronatus, 196
Philanthus crabroniformis, 13, 20, 23, 90–101; comparative nesting behavior, 231, 235, 240–243, 246, 247, 250; comparative reproductive behavior, 201, 205, 208–213, 218, 221, 222, 225, 227
Philanthus crotoniphilus, 31, 114
Philanthus gibbosus, 14, 20, 26, 78–90, 209; comparative nesting behavior, 231, 233, 235, 239, 242, 243, 247–251
Philanthus gibbosus group, 10, 78–114, 255
Philanthus inversus, 10, 22, 109, 112–114; comparative nesting behavior, 235, 242, 243–247, 250
Philanthus lepidus, 10, 158, 169–174; comparative nesting behavior, 235, 242, 246, 247, 250
Philanthus multimaculatus, 105–111; comparative nesting behavior, 232, 235, 242, 246–247, 250; comparative reproductive behavior, 209, 212, 217, 227; territoriality, 16–17, 147
Philanthus neomexicanus, 24, 138

Philanthus pacificus, 130–133, 148, 209; comparative nesting behavior, 235, 242, 243, 247, 250
Philanthus pacificus group, 10, 115–139, 255
Philanthus parkeri, 8, 164, 226, 227, 257
Philanthus politus, 148, 158–164, 213; comparative nesting behavior, 235, 242, 246, 247, 250
Philanthus politus group, 10, 140–164, 255
Philanthus psyche, 15, 140–152; comparative nesting behavior, 231, 232, 235, 242, 246–247, 250; comparative reproductive behavior, 200, 205, 208, 210–218, 221, 226, 227
Philanthus pulcher, 90–91, 115–130; comparative nesting behavior, 231, 232, 235, 242, 243, 246–248, 250; comparative reproductive behavior, 205, 209–215, 217–219, 221–222, 225–227
Philanthus pulcherrimus, 197
Philanthus sanbornii, 70–77; comparative nesting behavior, 232, 235, 238, 242, 246–247, 249–251; predation on honey bees, 23, 25, 31, 74–76
Philanthus serrulatae, 154–158, 167; comparative nesting behavior, 232, 233, 235, 242, 247; comparative reproductive behavior, 209, 212, 213, 218
Philanthus solivagus, 178–182; comparative nesting behavior, 235, 242, 246–247, 250, 256
Philanthus tarsatus, 151–154, 209, 213, 227
Philanthus triangulum, 1, 8, 192–196; hunting behavior, 30, 195, 245; natural enemies, 26–28, 31, 195–196; nesting behavior, 194–195, 233, 235, 239; orientation, 20; territoriality and scent marking, 12–16, 193–194, 209
Philanthus ventilabris, 140, 165–169, 209, 256; comparative nesting behavior, 235, 239, 242, 247, 250, 252, 255, 256
Philanthus venustus, 196–197
Philanthus zebratus, 20, 29, 45, 57–70, 100; comparative nesting behavior, 232, 233, 235, 242, 247, 250; comparative reproductive behavior, 201, 205–206, 209, 210, 213, 214, 218–224, 258
Philanthus zebratus group, 10, 32–77, 255, 256
Phoridae, 182
Phrosinella, 30, 178, 234, 250; *P. fulvicornis,* 77, 89, 162, 182, 191; *P. pilosifrons,* 70, 101, 129, 138
Physocephala, 27, 195

Piek, T., 240
Podalonia, 42–43, 206, 210, 245, 249
Pompilidae, as prey, 55, 68, 151
Population size, variation in, 155, 232–233
Powell, J. A., 130–133
Predatory behavior. *See* Prey capture; Provisioning; Stinging
Prey capture, 22–25, 30–31, 239–240; by *Philanthus albopilosus*, 189; by *P. bicinctus*, 41; by *P. crabroniformis*, 97; by *P. gibbosus*, 87; by *P. multimaculatus*, 111; by *P. sanbornii*, 74; by *P. triangulum*, 195
Prey carriage, 22–23, 240, 249
Proctacanthus micans, 38, 42
Protandry, 10, 58, 202–203, 259
Provisioning, 22–26, 243–247; by *Philanthus albopilosus*, 188–191; by *P. barbatus*, 104–105; by *P. barbiger*, 138–139; by *P. basilaris*, 54–56; by *P. bicinctus*, 41–42; by *P. bilunatus*, 177–178; by *P. crabroniformis*, 96–100; by *P. gibbosus*, 86–89; by *P. inversus*, 113; by *P. lepidus*, 173–174; by *P. multimaculatus*, 109–111; by *P. pacificus*, 132; by *P. politus*, 161–163; by *P. psyche*, 150; by *P. pulcher*, 125–129; by *P. sanbornii*, 73–75; by *P. serrulatae*, 157–158; by *P. solivagus*, 180–182; by *P. triangulum*, 195; by *P. ventilabris*, 168–169; by *P. zebratus*, 66–67
Pseudanthophilus, 255
Pseudoscolia, 3, 254

Rathmayer, W., 2, 23–24, 192
Rau, P., 78, 89, 165–167
Receptivity of females, 201, 259
Reinhard, E. G., 18, 78–82, 86, 87, 89
Removal experiments, 7, 221–222; on *Philanthus barbiger*, 136; on *P. basilaris*, 48–50; on *P. crabroniformis*, 94; on *P. psyche*, 143; on *P. tarsatus*, 154
Reproductive success: of females, 248; of males, 220–226
Ristich, S. S., 178, 180, 182
Robber flies. *See* Asilidae

Sand wasps. *See* Nyssoninae
Satellite flies, 28–29, 241–243. See also *Hilarella hilarella*; *Senotainia*
Satellite males, 17, 18, 223–226
Scelionidae, as prey, 126
Sceliphron assimile, 203
Scent marking, 12–13, 211–215, 217, 257, 258; evolution of, 214–215, 253; rates of, 212–214. *See also* Territoriality
Schmidt, J. O., 9, 13, 15, 16, 33, 37, 215
Seasonal patterns of activity, 10, 202–203, 231–232
Seger, J., 71, 75, 77
Senotainia, 28–29, 86, 110, 111, 177, 250
Senotainia rufiventris, 150
Senotainia trilineata, 242–243, 249; parasitic on: *Philanthus barbiger*, 138; *P. basilaris*, 57; *P. bicinctus*, 44; *P. bilunatus*, 178; *P. crabroniformis*, 101; *P. gibbosus*, 89; *P. inversus*, 113–114; *P. lepidus*, 174; *P. pacificus*, 133; *P. politus*, 162; *P. pulcher*, 129–130; *P. sanbornii*, 75–77; *P. solivagus*, 182; *P. zebratus*, 29, 70
Senotainia vigilans, 162, 174, 249
Sex ratio, 11, 203, 259
Sexual dimorphism, 8–10, 32, 202
Sexual selection, 198–199, 228–229
Shappirio, D. G., 90
Simonthomas, R. T., 2, 6, 11, 192–196, 233; on nesting behavior, 24–25, 28, 44; on reproductive behavior, 16–18
Size, importance of: in females, 129, 220; in males, 18, 220–226, 259; in males of *Philanthus basilaris*, 48–49, 52; in males of *P. crabroniformis*, 94; in males of *P. psyche*, 143; in males of *P. pulcher*, 120–121; in males of *P. zebratus*, 61–64
Size variation, sources of, 217–220
Spatial distribution, 203–211, 231–232
Sphaerophthalma, 90
Sphecidae, classification of, 2–4
Sphecinae, 4, 239, 257
Sphecius, 5, 208; *S. grandis*, 208, 248; *S. speciosus*, 198, 201, 204, 208, 214, 217, 249
Sphex, 231, 237
Spiders, as predators, 26, 43, 70, 90, 150
Spofford, M. G., 248, 249
Steiner, A., 239–240
Steniolia obliqua, 204, 206
Stinging behavior, 23–24, 239–240
Stubblefield, W., 24, 71, 75, 164, 226

Tachysphex, 202, 208, 214, 249
Tachytes, 200, 206, 209
Territoriality, 11–13, 16–17, 204–227, 256, 258; of *Philanthus barbatus*, 101–103; of *P. barbiger*, 134–136; of *P. basilaris*, 46–53; of *P. bicinctus*, 34–38; of *P. bilunatus*, 175–176; of *P. crabroniformis*, 92–94; of *P. gibbosus*, 80; of *P. multimaculatus*, 106–108; of *P. pacificus*, 131; of *P. politus*, 159; of *P.

Territoriality (*continued*)
  psyche, 141–147; of *P. pulcher,* 117–123; of *P. sanbornii,* 71–72; of *P. serrulatae,* 155–156; of *P. tarsatus,* 152–154; of *P. triangulum,* 193–194; of *P. ventilabris,* 166; of *P. zebratus,* 61–62, 64
Thermal constraints, 145–146, 227
Thiem, H., 30
Thornhill, R., 198, 199, 203, 215, 223
Tinbergen, N., 1–2, 20, 23, 24, 192, 194, 233
Tiphiidae, as prey, 55, 76, 105, 139, 169
Tracheloides, 254
Trachypus, 2–3, 239
Trigonopsis cameronii, 201, 204, 208
Trivers, R. L., 198, 200

Trypargilum, 199–201, 208, 214, 218, 221, 226
Tsuneki, K., 196, 227, 237

Velvet ants. *See* Mutillidae
Venoms, 24, 240
Vinson, S. B., 212

Wcislo, W. T., 232, 237
West-Eberhard, M. J., 253
Wheeler, W. M., 198
Willmer, P. G., 220, 227, 248
Wind, effect on scent marking, 144–145, 213–214

Zodion, 27, 70, 130, 164